Fast Robots:
Building Flexibility

Molina

Contents

Chapter 1

Introduction

Legged motion has been idealised as the best method for locomotion on land as it allows for passage through rugged terrain and a range of rapid manoeuvres. Manoeuvrability is also paramount to survival in nature seen through the agile motion such as rapid acceleration, abrupt stops, running jumps and near-instantaneous turns that seem effortless to legged animals [4][5][6]. However, there is minimal research investigating how animals and humans are able to perform these agile motions. Figure 1.1 shows an example of such a manoeuvre by Usain Bolt at the start of his 100m sprint.

Figure 1.1: Usain Bolt at the start of the 100m sprint [7].

1.1 Aims of Project

The aim of this study is to build a bipedal robot suitable for rapid acceleration manoeuvres. This project will focus on investigating existing platforms and use the information gained to design and optimise a new bipedal robot with high agility. Furthermore, the supporting hardware, software and a simplistic control strategy were implemented to allow the robot's performance to be verified. This platform will be used as a first prototype from which further iterations should be designed and novel rapid acceleration controller templates tested. This robot is named *Baleka* which comes from the isiXhosa word for run.

1.2 Motivation

The aforementioned transient motions are still not very well understood by biologists and most studies have taken inspiration from nature. The successful morphology for animals is due to natural selection which has allowed an optimal design to be found given the numerous trade off's for such manoeuvrability. Numerous legged robotic platforms are part of this trending field with exceptional robots created such as the MIT Cheetah [8] inspired by the Cheetah, the KAISTs Raptor 2 [9] inspired by the Raptor, FastRunner [10] inspired by an Ostrich and many others [1][11][12]. However, the controllers for these platforms are designed to achieve similar steady state locomotion characteristics as their biological counterparts with none having studied the aperiodicity of rapid acceleration manoeuvres.

Robots and animals are fundamentally different, where robots comprise of stiff rigid links and rotary actuators, while animals are compliant with muscle tendon actuators [13]. Thus, the morphology for an animal may not be optimal for a similar bio-inspired robotic platform and may be detrimental for rapid manoeuvres.

Henceforth to design such a bipedal robot, the rapid acceleration of such a model must be studied of which there currently exists no prior research. As of writing there exists only several papers covering rapid acceleration investigations with one concerning a bipedal humanoid model [14][15][16].

The purpose of designing a robot optimised for rapid acceleration motions is to allow future students to test and verify novel controller templates on a physical platform which was not previously available at the University of Cape Town. The reader should note that a template is a simplified model of a complex system with similar characteristics (explained further in Section 2.6)

1.3   Objectives and contributions

This project is broken down into three sections, namely: the design and optimisation of the robotic leg, the controller design and lastly the verification of the robot's performance. The main objective is to design a functioning agile bipedal robot. This includes the required embedded system, hardware and real-time operating system. There are numerous sub-objectives required to achieve this main objective.

A summary of the sub-objectives for the mechanical design of the robotic platform:

- Review of existing robots to select a suitable leg topology and actuator scheme;
- Gain insight from trajectory optimisation for mechanical parameters of a rapidly accelerating robot, and;
- Design, manufacture and assemble the robot.

Furthermore, for the control and verification aspect:

- Model the dynamics of the bipedal robot using the CAD model;
- Design and simulate a simplistic hopping control strategy;
- Develop supporting hardware and software infrastructure, and;
- Perform verification experiments on robot platform using the controller.

1.4   Scope, Limitations and Constrains

There were numerous constrains and limitations associated with this project that had to be considered during all design choices. This project was constrained by:

- Pre-bought motors and motor drivers;
- Available and affordable materials for machining;
- Pre-determined supplier for gearboxes for the drive train (Matex), and;
- Project time (24 months).

To ensure that the platform was completed within the 24 months, several limitations were introduced. Investigations of rapid acceleration manoeuvres were limited to flat known environments, thus the robot did not need to be optimised for ragged terrain.

The novel controllers that will be implemented on the robot by future students are expected to be limited to the sagittal plane. It is thus outside the scope of this project to design a robot for full three dimensional motion, limiting the robots required motion to the sagittal plane. Designing a novel controller template for rapid acceleration to control the bipedal platform is also outside of the scope of this project and therefore an existing control methodology was designed for performance testing.

1.5   Project Outline and Contributions

The primary aim of this book is to build a highly agile bipedal robot. This book is structured such that the reader can follow the investigations and decisions made throughout the project. Figure 1.2 provides a high level overview of this book. A novel contribution was made entitled, *Investigation of a Bipedal Platform for Rapid Acceleration and Braking Manoeuvres*. Trajectory optimisation was used to assist in the mechanical design of the bipedal robot for rapid acceleration manoeuvres and has been published in the top ranked conference on robotics, the International Conference on Robotics and Automation [17][18].

The flow and structure of the book is shown in Figure 1.2 which can be grouped into four categories. The Preamble presents existing agile robots and how they achieved it. The Leg Design Realisation describes how the author arrived at the physical robot design. The Controller and Infrastructure presents the controller design for the bipedal robot and the supporting systems to run the robot. Finally the assembled robot is experimentally tested and evaluated.

The book begins with the first and second chapter providing the reader with a background of agile legged locomotion, common design trade-off's and a detailed review of existing legged robots. The third chapter clarifies the project requirements, expands on constraints and limitations and describes the methods taken to achieve the final aim.

Chapter 4 evaluates the existing leg topology and actuation strategies. By analysing the force transformation properties of the leg linkages, a suitable topology was chosen. Furthermore, a selection process identifies the most suitable actuation method for agile manoeuvres given the constraints of the project. By using optimisation techniques, the trajectories for rapid ac-

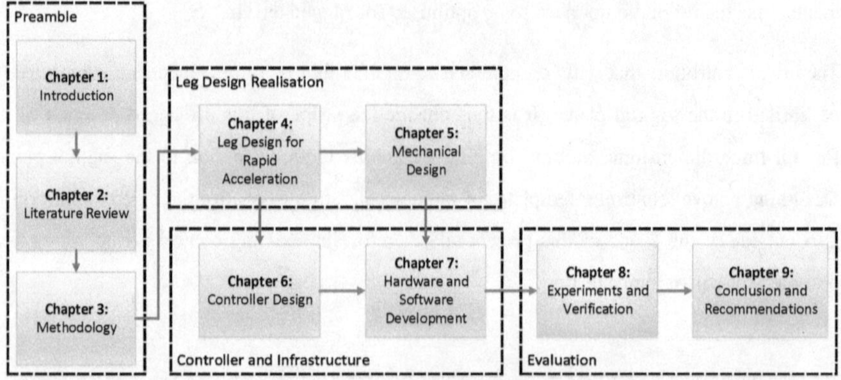

Figure 1.2: The flow and structure of the book chapters, grouped into four main categories.

celeration manoeuvres are generated, assisting in selection of physical parameters such as leg length.

Chapter 5 describes the iterative detailed design process to arrive at the final biped design. From the literature review, several vital requirements of the mechanical design were identified that improve the agility of robots. These requirements are closely followed through the numerous design iterations. Finite element analyses was performed on critical components to ensure a rigid, robust robot.

Chapter 6, a simplistic controller for the robot was investigated and designed. This controller was tested on a model of the robot in a physics engine to verify it's functionality.

Chapter 7 describes the hardware and software required to physically run and test robot. This concerns designing a boom to limit the robot to the sagittal plane, developing the real-time control system and the sensors required for the controller.

Chapter 8 evaluates the performance of the robot. The leg was put through incremental tests to ensure all supplementary systems were functioning correctly. This was followed by agility tests using a known metric to enable comparisons with existing robots. Lastly, the physical robustness of the robot was investigated.

Chapter 9 completes the book by evaluating to what extent the robot meets the project require-ments set out in Chapter 3. Recommendations on future improvements are then described.

Chapter 2

Literature Review

This chapter serves to analyse the current legged platforms and provide the groundwork for the leg morphology and actuator scheme chosen. This will also cover the compromises that had to be made between the desirable properties of highly manoeuvrable robots. The information gathered here led to the design of Baleka the biped robot. The various robots compared can be seen in Figure 2.1. In the section below, a quick overview is given of the notable properties of several existing robots. There are several terms that may be unfamiliar to the reader but are extensively covered further on in this chapter.

2.1   Summary of existing robots

The first dynamic robotic platforms to be developed were Marc Raibert's self-balancing dynamic hopping robots [3]. These robots were proficient at steady state locomotion and responding to external disturbances but were not designed with manoeuvrability as a focus. His robots used a prismatic leg that decoupled the leg length and leg angle for simplified control. These robots had to use pneumatics which requires significant infrastructure (air compressors) but has a very low control bandwidth.

The design prioritisation of the MIT Cheetah was to build a dynamic robot that was bio-inspired by the Cheetah and was capable of high velocity bounding locomotion [26]. It was built with custom high density, low impedance actuators allowing for great force transparency [27][28]. It used a series articulate leg linkage which was constrained to the sagittal plane and performed optimally for forward locomotion but restricted vertical jumping and backwards running.

The GOAT (Gearless Omni-directional Acceleration-vectoring Topology Leg) made use of a direct drive transmission with high density rotary actuators allowing for high control bandwidth

Figure 2.1: Existing legged robotic platforms along with the type of linkage mechanism shown below adapted from [19]. Starting top left to bottom right: StarlETH [20], Minitaur [21], MIT Cheetah [22], Big Dog [23], GOAT leg [24], Raibert 3D hopper [25] and ATRIAS [1]

and force transparency similar to the MIT cheetah. The leg uses a linkage mechanism with three motors and offers a three dimension workspace enabling 3D jumping [19].

The BigDog is a quadruped with redundantly articulated legs using a hydraulic pump to drive low friction hydraulic actuators [29]. This allowed for accurate foot positioning but the number of actuators located in the legs increased leg weight and resulted in poor high speed performance. The lower limbs also have compliance to mitigate some of the impact forces.

Ghost Minitaur is a quadruped using a scissor linkage mechanism driven by two direct drive actuators. The scissor linkage mechanism is symmetric and is suitable for a wide variety of tasks [21]. This platform is able to jump and run effectively with negligible leg mass as the actuators are located at the hip [30].

The StarlETH quadruped was designed for impact mitigation, forward locomotion and passage through ragged terrain [11]. It makes use of high geared motors to enable high torque output but mitigates ground impacts by using series elastic actuators (SEAs) [31]. In addition, SEA allows the robot to sense ground forces but limits the control bandwidth of the output.

The last platform investigated was ATRIAS, a bipedal robot that also makes use of a SEA to improve energy efficiency by storing and reusing energy from past gaits, similar to tendons in animals. This robot makes use of the same linkage as Minitaur but uses an extremely high gear ratio to account for the 62kg mass of the robot and suffers from low control bandwidth due to the SEA [1].

It is also worth mentioning legged robots with one dimensional legs. These leg mechanisms use a single motor and typically follow a set end affecter trajectory. These have been seen to transverse ragged terrain as seen with RHex [32] or run at speeds up to 40km/h as seen with the Rapter 2 [9]. Raptor 2 is only suitable for near steady state locomotion and RHex is more focused on moving through rough terrain than high-speed locomotion, making both these designs unsuitable for this project.

The design prioritisation of each legged robot mentioned above had varied significantly depending on the desired motion. This owes to the compromises between actuator schemes, leg topologies and control methods when designing a legged robot. The main compromises seen are [21]:

1. Torque density versus proprioceptive force sensing (high force transparency);
2. High speed legs versus leg strength (inertia) and robustness, and;
3. Passive compliance and energy efficiency versus controllability/control bandwidth.

This has resulted in a wide range of existing robots, emphasising the importance of evaluating key features for a highly agile bipedal robot. Several of these features that affect the agility seen in the robots mentioned above are expanded upon below. Those being high powered legs, compliance within the leg, platform robustness and proprioception. This was followed by investigations of the existing transmission methods, actuators and leg topologies.

2.2 Highly Agile Leg Design

The leg morphology greatly affects how the robot performs and needs to be chosen carefully to maximise the desirable features. This section covers the main characteristics seen in the existing platforms along with the relevant advantages and downfalls.

2.2.1 High Powered Legs

Force output: In many of the existing platforms, it is essential that the foot force is maximised as legged platforms typically experience extremely high ground reaction forces. The impulse due to gravity during the aerial phase of running has to equal the impulse during the foot contact stance. As the forward locomotion velocity increases, the stride frequency increases, thus reducing the foot contact time and increasing the required ground reaction force. In humans, it is seen to be around three times the body weight [33]. This is significantly less for quadrupeds seen at around 2.6 times the body weight of a dog [34].

There is a trade-off between stride frequency and force output for motors. However, in legged robots, it is most commonly seen that robots are limited by torque output and not speed limitations [35]. Therefore, it is desirable for a high mass specific force (the force output per kg robot mass). To achieve this high mass specific force output, some large platforms such as ATRIAS used a high gear ratio of 56:1 [1] and BigDog makes use of high pressured hydraulic pistons [23]. Other smaller platforms sacrifice torque output for other benefits discussed in Section 2.3 below. This includes lower leg impedance for greater force transparency.

High Speed: For high speed locomotion, the stride frequency of the leg increases drastically. To swing a leg back and forth during this motion requires high leg accelerations. Thus, it is highly desirable to locate the the mass of the leg near the hip and lower the total mass leading to reduced leg inertia. This not only reduces the torque requirements to accelerate the leg but also minimises the opposing torque on the body. In both cases it is seen experimentally that with a leg mass less than 10%, the robot can be modelled accurately as the well known SLIP model (Spring-Loaded Inverted Pendulum, see Section 2.6.1) which has a massless leg, drastically simplifying the controller [36]. ATRIAS is designed such that the leg mass is less than 5% [1] while the MIT Cheetah achieves a leg to body mass of less than 10% [36]. Achieving this is mostly done by selecting a suitable leg topology that minimises mass and locates the actuators (normally the most heaviest element) near the hip. The leg topologies of existing platforms are reviewed in detail in Section 2.4.

2.2.2   Leg compliance

Compliance is seen in all legged animals through their muscles, tendons and fat [37] and forms a basis for the template of walking and running, which is known as the SLIP model [38][39]. The SLIP model can be seen in Figure 2.2 and is described in more detail in Section 2.6.1. There are currently two different methods for achieving compliance, namely; active compliance and passive compliance. Active compliance is when a mathematical model is used to mimic a physical spring and damper. Passive compliance is the use of a physical component as a spring in the mechanical system, as seen in the SLIP model.

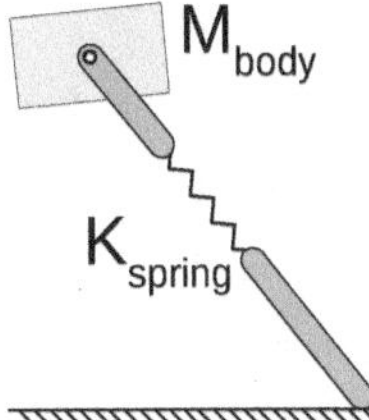

Figure 2.2: The standard Spring-Loaded Inverted Pendulum. The leg is massless and is represented by a spring attached to a body with a mass.

By using active compliance, the physical constants can be modulated real-time which is beneficial for the stability of the various gaits of legged locomotion. This however, does not allow the absorption and reuse of energy that passive compliance benefits from [40]. ATRIAS uses passive compliance and sees around 48% of the energy recycled from the previous step [1]. Passive compliance is typically implemented through the use of series elastic actuators (discussed further in Section 2.3) or direct springs as seen in the lower limbs of BigDog [23].

Passive compliance cannot be modulated and the elasticity is generally chosen around some predetermined motion, thus possibly detrimental to other manoeuvres [1]. Several attempts have been made to over come this with variable stiffness elements but this results in a mechanically complicated design prone to failure [41].

Active compliance comes with its own downfalls. The control system has to operate at near KHz bandwidths to simulate a physical spring [19]. In addition, to back-drive the motors, extremely low leg impedance is required (similar to how foot forces would compress a spring). Leg impedance is mainly caused by high gear ratios which increases the reflected inertia of the motor to the leg by a factor of N_g^2 (where N_g as the transmission ratio). Other physical properties that contribute to this are joint friction and damping. Thus, to use active compliance the linkage mechanism needs to be designed to minimise these affects and avoid gear trains greater than a factor of 10 [28]. However, an additional benefit of low impedance is proprioception, allowing the forces at the foot to be sensed at the motors. This is highly desirable for legged locomotion. Compliance in closely linked to the transmission methods and proprioception which are further discussed in Section 2.3 and Section 2.2.4 respectively.

2.2.3 Platform robustness

As discussed in Section 2.2.1, the ground reaction forces generated by legged platforms are extremely high. It is essential that the mechanical design is strong enough to withstand the impact forces over numerous cycles without the chance of fatigue failure. However, the trade off of robust legs is increased weight which, as previously mentioned, needs to be reduced. This indicates that high strength, low density materials are desirable for the mechanical design of a highly agile robotic platform. Some of the preferred materials are for example, titanium (Ti-6Al-4V) with a yield stress to density of 0.226 MPa/kgm^{-3} and heat treated aluminium (6083-T6) with a yield stress to density of 0.1 MPa/kgm^{-3}. The leg topology can also be selected to improve the robustness such as the scissor linkage used by Minitaur which breaks the ground reaction forces between two sets of links [21] (see Figure 2.6). MIT has also used tendon like features to drastically improve the strength of robotic legs while avoiding weight increases through biomimicry [8].

Mitigating impact forces are also a suitable method for increasing mechanical resilience. Most gear trains are not back drivable, but by using series elastic actuators the output linkage is decoupled from the drive transmission. Thus, impacts are absorbed into the compliant element avoiding high torques driving into the drive train [42]. Another method is to avoid the use of gear trains entirely reducing the mechanical impedance allowing backdrivability of the motors which has seen use in numerous robots such as Minitaur [21] and GOAT with direct drive [19] and the MIT Cheetah with a low gear ratio of 1:5.8 [27].

2.2.4 Proprioception

Proprioception is the visibility of forces at the foot. It is important for impedance control systems and most robots rely on force information from the end affecter to control the robot's motion. The most straight forward method is to use a force sensor on the end effector but the price of a three axis force sensor along with the high impacts of legged robotics makes this an impractical, fragile option. Boston dynamics BigDog makes use of a typical force sensor on its feet [23]. Another alternative is strain gauges. However, these are extremely susceptible to vibrations and accelerations that are common in mechanical linkage systems such as a robotic leg, ruling it out as a suitable force sensing method.

StarlEth and ATRIAS make use of series elastic actuators to determine the force at the foot by measuring the deflection of the spring with a known spring constant (explained in Section 2.3.2) [1][31]. Minitaur, GOAT and the MIT Cheetah use a technique called Generalized momentum based-observer which, through a model of the motor and linkage mechanism, can determine the force experienced by the end affecter by sensing the motor current [43]. Another force sensing technique was to use a rubber foot with numerous pressurised pockets of air with pressure sensors. Through machine learning algorithms, the MIT laboratory was able to determine the forces in all directions. However, this is highly specialised and not currently available to purchase [44].

From this it is seen that the most realistic approach to determining foot force is the use of torque (current) feedback. Thus the legged system should have the leg impedance minimised as far as possible.

2.3 Existing Actuation and Transmission Methods

There are numerous different transmission and actuation methods, several of which have seen use in the robots covered above. The methods to drive the robot legs have numerous trade-offs that need to be selected such that it is optimised for rapid acceleration. This decision will be made in Section 4.1. Herein the different pitfalls and advantages are reviewed. The different transmissions can be seen in Figure 2.3.

2.3.1 Gear Train

The major advantage of a high transmission gearbox is the torque density increase from the motor output. This can be hundreds of times larger than the motor's standard torque output. Nevertheless, geared motors cannot be back-driven due to the effects of reflected inertia. For instance, if a motor had a inertia of I_m, a torque of T_m, a speed of ω_m, a gearbox transmission ratio N, an output inertia of I_L, output torque of T_L and speed of ω_L. It is known that:

$$T_m \omega_m = T_L \omega_L \tag{2.1}$$

$$T_m N = T_L \tag{2.2}$$

$$\frac{1}{N} \omega_m = \omega_L \tag{2.3}$$

The equation of motion for the motor is:

$$T_m = I_m \dot{\omega}_m \tag{2.4}$$

Substituting eq. 2.2 and eq. 2.3 into eq. 2.4 gives:

$$\frac{T_L}{N} = N I_m \dot{\omega}_L \tag{2.5}$$

$$T_L = N^2 I_m \dot{\omega}_L \tag{2.6}$$

According to equation 2.6, the inertia of the motor is increased N^2 from what is seen by the torque acting on the load. This drastically reduces the backdrivability of the motor. Thus, during a foot contact impulse, the external force is driven directly into the gearbox causing extremely high stresses on the gear teeth. This can be detrimental, resulting in reduced robustness [45]. In addition, geared motors suffer from low efficiencies and directional inefficiencies which depends on the direction of power flow, complicating control [46].

Several further detrimental characteristics are that of backlash, stiction and friction. Backlash is a discrete problem and has to be accounted for within the controller. This not only complicates the control system used, but often requires an additional encoder to determine the position and velocity of the output link [47]. Harmonic drives exist that offer no backlash but are drastically more expensive and still suffer from high reflected inertia. The above factors drastically decrease the motor's bandwidth and thus a geared motor is highly undesirable for agile manoeuvrability and are not used alone in any of the robots reviewed in Section 2.1.

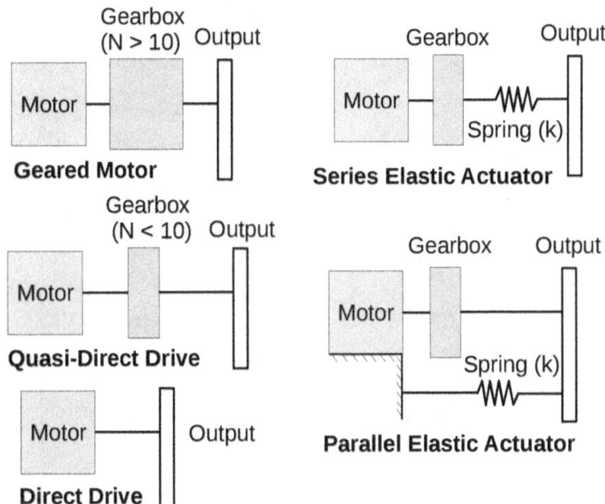

Figure 2.3: The various different transmission types, from top left to bottom right: geared motor (GM), Series Elastic Actuator (SEA), Quasi-Direct Drive (QDD), Direct Drive (DD) and Parallel Elastic Actuator (PEA).

2.3.2 Series Elastic Actuators

A series elastic actuator (SEA) is where a spring is placed in series between the actuator and the output. This compliant element is most typically found between the gearbox and output link. There are several advantages to using a SEA however, there are significant design trade-offs.

SEA's act as low pass filters for any shock impulses, reducing the force peak that often occurs during ground contact in legged robots, thereby improving the robustness [42].

Impedance control is the main output for controllers of legged robots, however, measuring this output is near impossible with a geared motor. SEAs offer proprioceptive force sensing by turning the force control problem in a position control problem [42]. Position is typically easy to control through a gear train, thus the output torque is easy to control [31]. The torque is calculated by using Hooke's law and knowing the spring stiffness and spring deflection (Eq. 2.7). For higher force sensing accuracy, the series elasticity K should be reduced.

$$T_{out} = K(\theta_{GearBoxOutput} - \theta_{output}) \qquad (2.7)$$

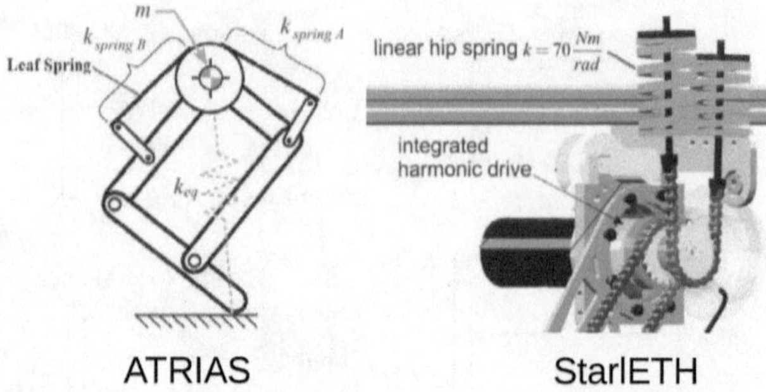

Figure 2.4: The ATRIAS leaf spring connected between the upper leg link and the motor. StarlETH with fixed linear springs connected via a chain to the motor.

However, the low pass filter property to avoid impacts also filters the motor outputs reducing the affective control bandwidth. The control bandwidth is inversely proportional to the spring stiffness and by increasing stiffness, the force measurement accuracy is reduced but the control bandwidth achievable will be greater and visa versa [42]. Low control bandwidth is highly undesirable for rapid manoeuvres as motions need to happen very quickly and accurately. It is seen that SEA's control bandwidth is limited to about 28Hz for small amplitudes and 11Hz at saturation [48].

A spring is a passive mechanical device that can absorb energy in the form of potential energy and release it back into the system again. This has been used to recycle gait energy in legged robots such as in ATRIAS where efficiency is improved by 60%. However the SEAs used in the ATRIAS had the spring stiffness chosen such that the passive dynamics resembles the target dynamics for the robot in steady state motion [1]. This is derived from the SLIP model which has been identified as a template for all forms of legged locomotion [38][39]. However, for different motions the optimal spring stiffness changes which limits the usefulness of SEA's [19], especially when the desired motion is not well understood at the time of design.

Some methods for designing the physical SEA are using leaf springs as in ATRIAS [1], ScarlEth [31] uses linear springs fixing the motor to the body and the Raibert hoppers [25] use the compliant nature of pressurised air. The SEA's for ATRIAS and StarlEth can be seen in Figure 2.4.

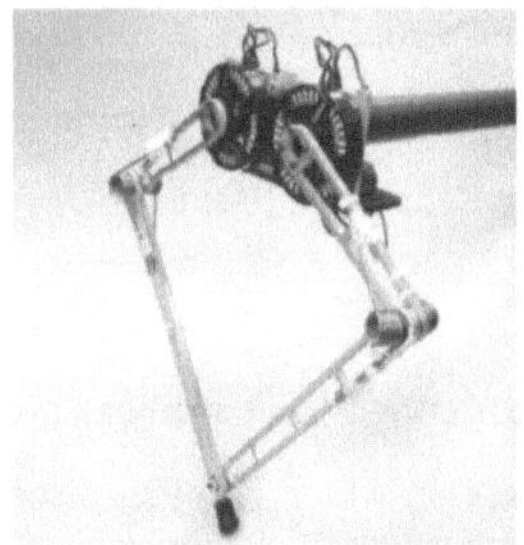

Figure 2.5: Parallel elastic actuator located at the joints of a scissor linakge for a monopod [50].

2.3.3 Parallel Elastic Actuators

Parallel elastic actuators (PEA) are where a spring is located in parallel with the motor and the output as seen in Figure 2.3. PEA's offer the benefits of storing and releasing energy like the SEA. However, the main difference is the ability of the PEA's to increase the possible torque output by outputting torque additively with the motor, decreasing the torque requirements for the motor [49]. An example of a monopod with PEA's can be seen in Figure 2.5 [50]. Oppositely, when resisting an external torque, the motor only has to provide a portion of the torque in resistance and the spring will absorb the rest [51]. PEA's do not affect the bandwidth of open loop control as the spring does not filter the output from the input.

This intuitively has uses in legged locomotion where, during launch, spring energy can be used additively with the motors to propel the robot. During landing the spring can store energy. The PEA forgoes the ability to mitigate the impact like a SEA does. PEA's are most suitable for direct drive or quasi-direct drive transmissions to enable backdrivability. They can also be used with a SEA such as in the work investigated by Yesilevskiy [49].

2.3.4 Direct Drive

Direct-drive is where the actuator is connected directly to the output (see Figure 2.3). Using direct drive requires high density actuators as there is no gearing to increase the torque output. This can cause higher energy joule losses in the motor as the operating range for electromagnetic motors is at higher speeds and lower torques.

Direct drive provides high force transparency as there are no gears between the actuator and the output and further forgoes the negative affects of gearing mentioned in Section 2.3.1. The affective control bandwidth is in the KHz range [21] and is perfect for active compliance (see Section 2.2.2).

The force at the foot is transferred through the leg linkage and directly into the air gap of the actuator, allowing for proprioception which eliminates the need for torque/force sensors or SEA's [27]. The force at the foot can be calculated accurately given minimal stiction and friction in the linkage mechanism. The current in a motor's windings (I_{im}) is multiply by the torque constant (K_T) to find the torque (T_{im}) and using the Jacobian transpose inverse (J^T), the force vector (F) at the end affecter can be determined, as seen below [52]:.

$$T_{im} = K_T I_{im} \tag{2.8}$$

$$\mathbf{F} = (J^T)^{-1}\mathbf{T_m} \tag{2.9}$$

The Jacobian matrix (J) is determined by the partial derivative with respect to the motor coordinates:

$$J = \frac{\partial \mathbf{X}}{\partial \mathbf{q}} \tag{2.10}$$

Where $\mathbf{X} = f(\mathbf{q})$ is the forward kinematics equation and $\mathbf{q}$ is the motor coordinates (the Jacobian can be further used to also determine the velocities of the end affector, $\dot{\mathbf{X}} = J\dot{\mathbf{q}}$). Furthermore, this reduced leg impedance also allows external forces to easily effect the motor dynamics improving the mechanical robustness by mitigating impulses that would normally occur on gear teeth providing a similar benefit to that of SEA's.

Given that the only downfall is torque density, numerous robots are moving towards direct-drive as it has such great controllability and robustness. Numerous platform such as Minitaur and GOAT use direct drive and are able to perform extremely agile manoeuvres [19][21][50].

2.3.5 Quasi Direct-Drive

Quasi direct-drive (QDD) is similar to direct-drive but the torque output from the actuators is increased at a small cost to impedance, actuation bandwidth and reflected inertia. QDD is a gear ratio of less than 1:10 using a single stage planetary gear [27]. A gear ratio above ten typically requires a second stage and this adds issues similar to those of the geared motor and causes cascading gear interface losses and additional backlash [53][12]. Kalouche did extensive experiments to compare DD and QDD and found that at a small cost to actuation bandwidth and reflected inertia, the quasi-direct drive can provide significantly more torque which is highly beneficial if rapid acceleration manoeuvres are required [19]. The MIT Cheetah saw this trade off and used a gear ratio of 5.8:1. MIT also saw that their use of QDD maintained a proprioceptive torque sensing with less than 5% error [27].

2.3.6 Hydraulics and Pneumatics

Pneumatics were first used in legged motion in the original Raibert hoppers [3]. Using pneumatics pistons to extend and withdraw the leg proved to be very useful as it decoupled the angle and length of the leg (using the SLIP model). Pneumatic pistons have high force density, low impedance and benefit from natural compliance due to the compressibility of air. This however is detrimental to the control bandwidth [54][55].

Conversely, no compliance is possible with hydraulics as fluid is incompressible, but results in high bandwidth control [55]. Furthermore, hydraulics are operated at high pressures and have a much higher force density than pneumatics. High pressures in excess 30MPa along with being flammable make it extremely hazardous [54]. Hydraulics also require more infrastructure than pneumatics as return tubes for the fluid are required compared to pneumatic cylinders which can just be exhausted into the atmosphere.

Downfalls for both are the requirement for auxiliary equipment such as pumps and pressure cylinders to operate. This makes it costly and heavy but has seen use in some well funded robots such as BigDog [29] and WildCat [56], both developed by Boston Dynamics.

2.4 Leg Topology designs

Selecting a suitable leg topology is a critical design choice that is inherently linked to a robot's overall motion [13]. In the past, the leg topology is designed such that spatio-temporal gait characteristics are suitable to perform a desired task [57]. However, for this project it is difficult

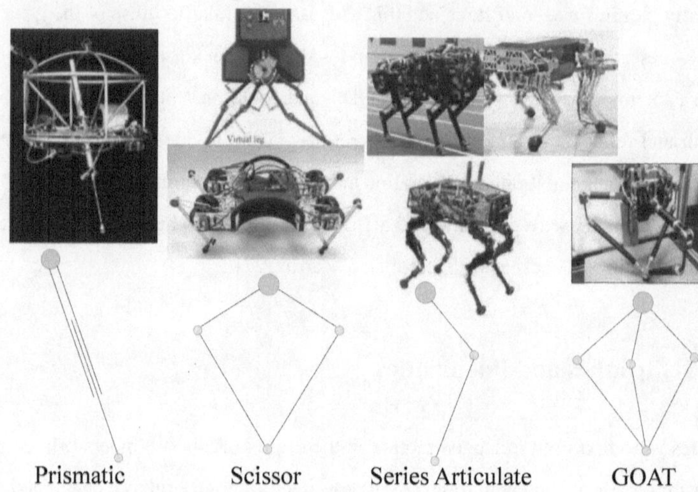

Figure 2.6: Existing legged robotic platforms along with the type of linkage mechanism shown below adapted from [19]. Starting left is the Raibert 3D hopper [25], ATRIAS [1], Minitaur [21], MIT Cheetah [22], StarlETH [20], Big Dog [23] and the GOAT leg [24]

to plan given the lack of research for rapid acceleration manoeuvres. The existing platforms reviewed in Section 2.2 are shown in Figure 2.6 with their respective linkage morphology.

2.4.1 Prismatic

The prismatic leg is the physical match of the SLIP model and decouples the leg angle and leg length. This mechanism is normally implemented by using a pneumatic cylinders, such as the prismatic leg seen in the Raibert Hoppers [3]. The pneumatics acts as a compliant spring to adsorb the ground impact. The angle of the leg is typically controlled with a geared motor due to the high torque requirements for single motor. However, this limits the accuracy at which the hip torque can be sensed (as explained in Section 2.3.1). As mentioned in Section 2.3.6, additional parts such as compressors make this design less feasible. SCHAFT has also used prismatic legs in the form of a drive train design, but this robot was designed for slow controlled motion and not manoeuvrability [58].

2.4.2 Series articulated

The series articulate leg has seen popularity in the later years with numerous robots using this design. This mechanism uses two motors to control the leg in a 2D frame. The series articulate mechanism does not effectively share the load between the motors throughout the work space [19]. This negatively effects the high force performance of the robot, specifically when large forces are demanded in unbalanced linkage configurations. However, the benefit that has drawn a lot of attention is the avoidance of negative work that sees a lot of energy wasted in closed kinematic chains such as the scissor leg mentioned below in Section 2.4.3 [1]. Since the design priority of numerous legged robots is energy efficiency, this linkage is selected.

There are two different variations that have mainly been developed. The MIT cheetah and StarlETH that make use of a linkage mechanism to locate both motors for the upper and lower links co-axially at the hip [36][31]. This reduces the leg inertia drastically and is suitable for rapid motions. The second variation is where hydraulic pistons are integrated into the legs and actuate the motion such as HyQ [55] and BigDog [23]. A last outdated design has one motor located at the knee which has to be swung with the leg (as in [59]), drastically increasing the leg inertia.

The series articulate configuration forgoes robustness as each link in the mechanism has to resist high bending moments mainly during ground contact impulses. Furthermore, the workspace for the series articulate linkage is asymmetric and this could affect the platform's performance during certain manoeuvres such as acceleration and de-acceleration.

2.4.3 Scissor Linkage

The Scissor linkage, similar to the series articulate linkage, has seen wide use in the legged robotics community. This linkage uses two motors in a closed kinematic chain (typically a four bar linkage mechanism) to control the position of the end affecter. This mechanism does not look like a typical leg found in biology but offers numerous advantages. The load sharing between the motors throughout the workspace is more balanced than the series articulate leg and thus, higher force demands can be met [21]. However, a portion of energy output by the motors is lost to negative work (called geometric power) that does not add to the force output

[1]. This was seen as a major downfall for ATRIAS whose design aim was to be a hyper efficient biped.

The Scissor linkage is significantly more robust as the ground reaction forces are distributed between two sets of links with the lower links only experiencing axial stress. An example of the minimal material required for such a mechanism, both the size of the ATRIAS leg and Minitaur's legs can be viewed in Figure 2.6 [30]. Both motors are located on the hip, further reducing the leg inertia. Most commonly the motors are located co-axially but can be located adjacently [50], as seen in Figure 2.5.

The workspace is symmetric which avoids biases for motions in different directions that the series articulate lnkage suffers from [30]. This can be desirable such as for the Minitaur robot whose design focus was a versatile robot with no foreknowledge of desired motions.

GOAT Leg

The GOAT leg linkage can be considered an extension of the Scissor linkage from 2D to 3D [19]. The end effector can be located in a 3D workspace by controlling the angle of three motors all located on the hip. Similar to the scissor linkage, it benefits from a symmetric workspace, better load sharing between motors and higher robustness as any ground reaction force is split between three different links. However, it also suffers from negative work due to the closed kinematic chain.

2.5 Bipedal Foot

The foot of a robot is a major factor in the motion achievable as it is the interface between the robot and the ground. It is desirable to have a high damping foot to avoid initial impact forces but a low spring property to avoid the foot ricocheting off the floor. Research has gone into an actuated ankle to improve running and walking [60]. Compliant feet have also been investigated concerning the mitigation of impact forces [61]. However, in numerous of the well known robots such as the MIT Cheetah and StarlETH, simple feet with rubber pads or high damping feet have been used [27][48].

2.6   Control Systems for Bipedal Robots

An initial investigation into the control systems that are used in bipedal robots was done.
Legged robots are typically extremely hard to control as they are under-actuated and are a
hybrid dynamic system with different dynamics for ground contacts and flight phases [3]. Templates are common with such difficult models to assist in the control. A template is a simplified,
generalized model whose dynamical properties are similar to the actual model. An anchor is
what links the template model to the physical constraints of the actual robot/model [62]. An
example of an anchor is the linkage system of a leg that anchors the spring-loaded inverted
pendulum (SLIP) template for legged locomotion (depicted in Figure 2.7).

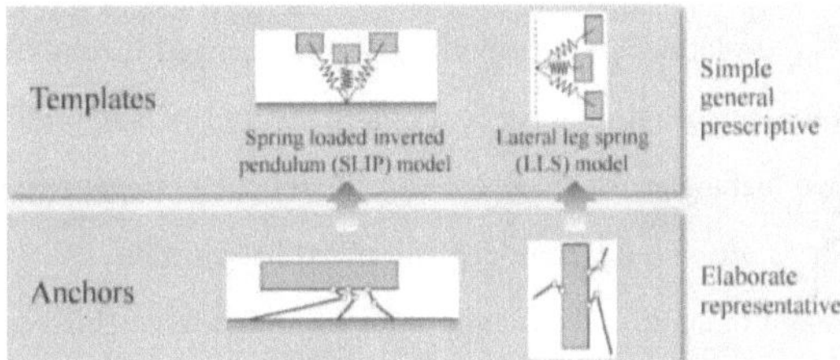

Figure 2.7: Visual interpretation of templates and anchors. The template is a simplified perspective/-
model. The Anchor, *anchors* the template to the physical world, such as modelling the kinematics of a
linkage mechanism [62].

2.6.1   Spring-loaded inverted pendulum

The Spring-loaded inverted pendulum has been identified as a suitable simplification of all
steady state legged robotic locomotion and has shown excellent performance in almost all
legged locomotion [38][39]. An image of this can be seen in Figure 2.2 and Figure 2.7.

ATRIAS uses a 3D locomotion controller that allows it to remain balanced by stepping from
right foot to left foot repeatedly. For its walking gait, the ATRIAS robot uses the SLIP model
and has a walking controller that alternates between the stance and swing phases for the legs.
It is designed such that each leg is out of phase from its counterpart to reduce the net torque
around the body [1].

Minitaur, even though a quadruped, has its dynamics ultimately reduced to the SLIP model, allowing for very agile, stable motions [63]. This is extremely useful as it simplifies the required controllers for legged locomotion.

The SLIP's spring element is either a passive element (such as a SEA) or controlled through active compliance. To determine the motor torques in active compliance, the desired spring force F is transformed into the joint space torques T_m by using the Jacboian transpose J^T as seen in equation 2.11 below.

$$\mathbf{T_m} = (J^T)\mathbf{F} \tag{2.11}$$

2.6.2 Legged Robot Controllers

The control of a legged platform is inherently difficult. Legged locomotion involves numerous collisions, requires precise timing for foot placement and falls into the category of hybrid dynamics. As the two feet make and break contact, the robot can have up to four different equations of motion. There are two different control areas that currently exists for legged robots. These exist in the realm of static and dynamic stability with a few hybrid control techniques to utilise both.

A statically stable robot is perfectly balanced without requiring movement (given no external forces) and can stay in this position indefinitely. This can be understood from the ground contact base that forms when a legged platform makes contact with the ground. To remain stable, the centre of gravity (CG) must remain within that base. For a quadruped, static stability is easily achieved as three of the feet must remain in contact with the ground at any time to for a contact triangle to be formed. This is easily visualised in Figure 2.8 (a). However, bipeds are typically unstable. To avoid toppling like an inverted pendulum, the CG must remain within the foot contact which is significantly smaller than the ground triangle formed in quadrupeds, making controller slower. Additionally, the torso must sway to cross the CG from one foot to the other, depicted in Figure 2.8 b). This type of control is known as the Zero-moment point (ZMP), where the summation of the torques generated by the ground reaction forces and body inertia always add to zero, avoiding a toppling moment that would cause the robot to fall over [64].

The control schemes utilised in static stability were based on position control, making motion

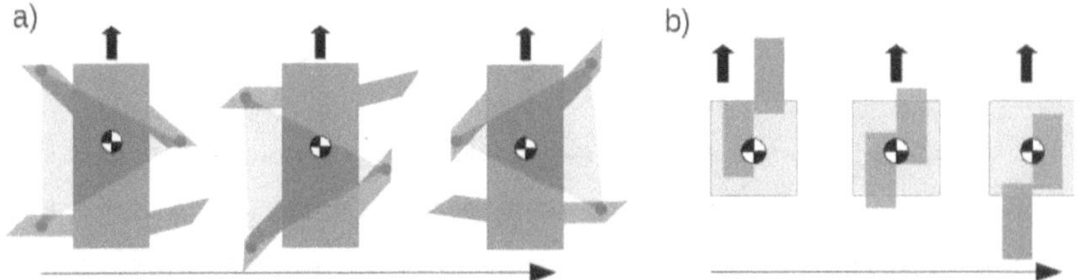

Figure 2.8: Ground contact base for legged platforms. Working from left to right shows forward movement in time and green indicates a foot/leg in the aerial phase. **a)** The CG remains within the triangular base that is formed by quadruped's feet during static locomotion. **b)** The ground contact area for the biped is only the singular foot requiring a large flat foot design. The biped's torso must also rotate to move the CG from one foot to the other during stance change (middle image).

extremely slow to ensure the limbs are placed precisely. Furthermore, this promoted heavy platforms with high gear ratios which is the exact opposite of the design focus for this project.

Dynamic stability was first introduced in Raibert's hoppers. These were reviewed in more detail in Section 2.2. Using simple calculations, these robots were able to achieve balance by constantly hopping like a spring, now defined as the Spring-Loaded Inverted Pendulum (SLIP) template mentioned above. Actively balanced systems can deviate from equilibrium and are designed such that any reaction in one direction is compensated by a reaction in the opposite direction [3]. The control architecture developed by Raibert was extremely robust such that it avoided the complexities of modelling the systems friction, inertias and masses entirely. This, back in the 1980s, was extremely useful due to the difficulties with predicting such values.

More research was completed by Raibert that explained how his simple monopod template and controller can be used on a bipedal platform. This involved only allowing a single leg to make contact with the ground at a time. Furthermore, the symmetry of locomotion is used to help balance the biped by commanding the legs to operate roughly out of phase, reducing the net torque acting on the robots body [65].

Further advanced adoptions of this original scheme has resulted in some very efficient (e.g. ATRIAS [1]) and agile (e.g. Minitaur [21]) robots. Other control schemes are not investigated given the limitations of time for this project. A simple robust hopping controller such as the Raibert controller is enough to verify the platforms performance and robustness. Note that the derivations for this controller are located in Chapter 6.

2.7   Trajectory Optimisation

In the past, during platform design, only simple calculations have been performed to identify the physical parameters for a task such as simple hopping. Furthermore, with complex manoeuvres, it becomes impossible to identify these parameters thorough basic calculations or simulations [13]. Trajectory optimisation is starting to close this gap. It is a typical optimal control problem however, it is formulated such that given numerous constraints, such as the equations of motion, a trajectory is generated from the decision variables that minimises the cost function. For the purpose of design, these decision variables can be applied to physical aspects of the robot. For more information see Kelly's tutorial [66]. With the improvements in computational power, trajectory optimisation has seen use in this manner, improving many different aspects of complex legged motion and assisting in design [67].

Numerous researchers have used optimal control to assist in their steady state (periodic) designs. This enabled engineers to only analyse a segment of movement, creating small-scale optimisation problems. Xi et al. used optimal control to minimise the cost of transport (COT) of a biped at a constant velocity (steady state) [68][69]. The gaits generated were purely the outcome of the optimiser and significantly improved the COT. This was evidence that the physical parameters could be derived from such a problem while aiming to optimise some output of the motion, in this case, COT.

On the other hand, Celik and Piazza investigated multi-step sprinting of a simple biped template [70]. They found similarities between the behaviours discovered by the optimiser and human motions during acceleration. One such motion is diving towards the end of the sprint to minimise time. Steenkamp took a step further and performed time optimal sprints for a quadruped and saw similarities between the optimal motions generated and those used by greyhounds [14]. The two aforementioned studies focused on investigating transient motion, however, neither focused on models for implementation on a robotic platform.

Optimal control can also close the gap between morphology and spatio-temporal gait characteristics by using realistic platform models. Khan et al. developed a method for actuator sizing of a quadruped robot given a desired jump height and trot velocity at different leg segment lengths [71]. Yesilevskiy et al. used optimal control to simultaneously optimize motion and spring positions (series and parallel elastic actuators) in a mono-pod hopper for minimum energy [49]. Spielberg et al. optimised for a simpler task requiring a robot to press a button, but allowed the optimiser to choose all aspects of the platform, from linkage lengths to actuator characteristics [13].

From the above it is seen that trajectory optimisation presents a method whereby motion too complex to calculate by hand can be investigated. Furthermore the literature shows that by cleverly specifying the problem, physical parameters for robots can be determined.

2.8 Summary

This chapter thoroughly reviewed the existing robots and their design focuses while arriving at the desirable traits for rapid acceleration. Such traits were proprioception, high robustness, high mass-specific torque and high speed legs. The various existing actuation methods, including the use of passive elements, were investigated with these characteristics in mind. Furthermore, existing linkage mechanisms were investigated and the various trade-off's were considered.

A simplistic controller suitable for legged locomotion was reviewed. In addition, trajectory optimisation was described as it could be used to investigate unknown motions and assist in the physical parameter selection for a robot. This Chapter provided the ground work for the author to begin the legged robot design process.

Chapter 3

Methodology

This project is broken up into four main areas that enabled the author to achieve an agile bipedal robot. These are:

- the leg topology selection;
- detailed mechanical design;
- hardware and software system design, and;
- the platform testing and verification.

This chapter defines the project requirements as well as the known constraints and limitations. This is followed by an outline of the methods and procedures taken to design the robotic platform, prepare the relevant hardware and software systems and ultimately bring all these together to verify the performance of the robotic platform.

3.1   Project Requirements Identification

The start of the design process was to investigate all the relevant literature on legged robots, mainly highly agile platforms. When designing a leg topology, it is important to decide which factors need to be prioritized. For most of the design decisions, the improvements of one characteristic results in the degradation of another. Concerning highly dynamic legged robots the following design characteristics are favourable for this project's design [19].

(a) High force/speed actuators;

(b) High control bandwidth;

(c) High proprioceptive sensing;

(d) Low impedance and limb inertia;

(e) Low body mass;

Table 3.1: Bipedal robot requirements for agile motion with the importance of each.

Req. No.	Description	Importance /10
1	Need a low mass/inertia leg to allow for rapid leg swings	10
2	Forces at the foot must be measured with some degree of accurately.	4
3	Leg impedance must be minimised for force transparency and backdrivability	8
4	The control system must have a high closed loop bandwidth	7
5	The mechanical design must be durable and robust	9
6	The robot controller must provide stable motions	6
7	Module leg design to allow the robotic legs to repositioned	10
8	Cost effective design to allow for future iterations	7
9	Allow for additional elements to be added such as springs, an actuated tail or various foot contact methods	4

    (f) High robustness, and;

    (g) Leg topology with a large workspace and force envelope.

At the start of the design process, it was important to identify the project requirements that are desirable from the final robotic platform. At the end of the project, the performance of the robot was compared against these requirements. The requirements of the project are written to be as independent of the design as possible. This was to prevent an initial bias when designing the platform. With further research the requirements were identified and assigned a value of importance to achieve the final goal, *a highly agile robot suitable for rapid acceleration manoeuvres.* These can be seen in Table 3.1.

However, there are also requirements of the research group that need to be taken into account when designing this platform. It was their desire and the author's to produce the first version of a legged platform that will continue to be used or modified for the next decade. These requirements are numbered 7 to 9.

3.2   Limitations and Constraints

There were several limitations and constraints associated with this project that had to be considered throughout the design process.

Proprietary motors (U12 T-motors) were pre-purchased along with the motor driver boards

(Ingenius Jupiter Drives). These two constraints guided the robot size as the maximum torque output was known.

The other main constraint was time. Most design procedures would have at least two functioning prototypes. However, this was not followed because this was the first robot to operate in the UCT Mechatronics lab and all supporting systems had to be set up in addition to the robot build.

The platform design and operation was limited to the sagittal plane (2D environment). Students in the Mechatronics Research Group are currently only investigating acceleration motions in two dimensions and thus the robot was not required to operate out of plane. The ground on which the platform ran was also limited to a flat known surface as again, the controllers that will be tested on the robot in the future are not concerned with ragged terrain.

3.3 Leg Topology Design

3.3.1 Transmission selection

With all the literature investigated and a set of requirements and constraints/limitations identified, the optimal actuator scheme had to be selected first. The actuator scheme greatly affects the performance characteristics of the robot and was ideal as the first step in design process [13]. Should the torque density be too low, the robot wouldn't be able to leap effectively. However, should the torque density be too high, the reflected inertia from the gearbox would be detrimental to the control bandwidth achievable. Thus, an actuator scheme was carefully selected that finds a balance between torque density, force transparency/proprioception, actuator bandwidth, robustness and mechanical simplicity optimal for highly agile motion.

3.3.2 Leg Linkage and Ratio

The next step in the design process was to select a suitable leg topology. Similar to the actuator, a linkage mechanism has a drastic affect on a legged robot's motion and force sensing capabilities. The serial articulate linkage, four-bar scissor linkage and the five-bar scissor linkage are investigated. These are specifically chosen as they have been implemented in some of the most agile robots that currently exist such as Minitaur and the MIT Cheetah [21] [27].

The linkage workspace, torque to force mapping from the motor to foot, linkage ratio and load distribution are the main aspects used to assist in the selection process. This section requires determining the forward kinematics of each linkage and using singular value decomposition (SVD). The singular values (a function of the geometry of the linkages) describes how each linkage performs [21]. The outcome was the selection of a leg linkage and ratio.

3.3.3 Trajectory Optimisation

Selecting a suitable nominal linkage length and gear ratio given the constraint of pre-bought motors was a challenge. For legged locomotion, the same forward velocity can be achieved by having a long leg and low stride frequency or visa versa. However, there are no calculations that can assist in the identification of the leg length, especially for the aperiodicity and complexity of rapid acceleration manoeuvres. Thus, a trajectory optimisation problem was created that minimises the time for a model's sprint over a known distance. The robot was constrained to start and end in rests, forcing an acceleration and braking phase to minimise time.

The problem was run iteratively, adjusting the leg length and gear ratio of the chosen linkage to find the model with the shortest sprint time. This involved calculating the Lagrange dynamics for the chosen linkage, deriving the optimisation constraints and starting conditions. This yielded the most suitable linkage and gear ratio for the robot to perform rapid acceleration and braking with the pre-purchased motors.

3.4 Mechanical Design

With the desired linkage selected and sizing determined, a detailed mechanical design was produced, going through several iterations and a prototype. Several embedded system parts were known and taken into account for the design. Furthermore, the leg design was kept modular as specified in the requirements so that it can be repositioned should it be desired. Several experienced engineers reviewed the mechanical design on numerous occasions providing useful suggestions to the author.

The prototype was built to take corrective action for any oversights. This concerns problems such as unexpected play and assembly issues. However, it was not a functioning prototype as it was created using 3D printed parts.

A second iterative process was required where critical components in the final robot design underwent stress analyses and modified to meet a suitable factor of safety. The forces used were determined from the ground reaction forces generated in the trajectory optimisation problem. The detailed mechanical component drawings were created and sent for manufacturing.

The quality of several components were substandard and had to be returned to the workshop for correction. When all the components were acquired, the modular legs were assembled.

3.5 Controller Design

From the literature review, the Raibert hopping controller was selected and designed (initially with a monopod) [3]. To verify the robustness of the platform and meet one of the projects requirements, designing a Raibert hopping controller would expose the platform to shock loads that can be expected during rapid acceleration manoeuvres. The sequential state machine of the monopod was then modified to control a biped, coupling the motion of each leg for a more stable system [65]. Both controllers were implemented in a robot simulator V-Rep which uses the Bullet physics engine. Here, the gains in the simulation were used as a starting point for when the controller was tested on the physical robot.

3.6 Embedded System, Boom and Control Environment

To test the robot, several systems had to be set up. The controller environment was chosen to be a Simulink Real-time target computer. It provides the excessive processing power of an Intel i5 CPU enabling even highly complicated controllers to operate at up to 1000Hz.

A communication protocol called EtherCAT was used to communicate with all other devices. EtherCAT is becoming increasingly popularity [72] and connects devices in a daisy chain. This was highly beneficial as a normal micro-controller would lack the computational power of the Intel CPU and would require communications to be configured individually for every device used.

The Simulink-Realtime system relies on the controller to be designed in Simulink and then compiled and pushed onto the target PC as a C++ program that runs at real-time.

Given the limitation of the robot to a 2D enviroment as stipulated in Section 3.1, a supporting structure in the form of a boom was required. Several different existing designs were compared. Some of the desired requirements were that it should be light weight, rigid, have a constant added mass affect on the robot and be mechanically simple. The best design was chosen and constructed.

The controller design reveals the required input data and thus, the sensors required for the robot. It was determined that the inertial frame position and velocity coordinates were needed. It was originally hoped that encoders on the boom could determine this, however, the length of the boom required an encoder with extremely high points per revolution to get a meaningful velocity reading. Conveniently, there was a motion capture system [1] located within the same lab (a local version of the Vicon system) [73]. This has sub centimetre accuracy and was modified to determine the position and velocity of the robot. Furthermore, to determine the attitude of the robot body, an encoder was incorporated into the design of the boom.

A method for detecting ground contacts was required to switch the controller through the different states. Two force sensors (ATI Axia 80F/T Sensor) were purchased to enable this switching state to be detected. A ground force plate was then created with these two sensors. Furthermore, proprioceptive capabilities of the robot can be determined by measuring the ground reaction force at the end affecter.

The Simulink-Real-time target computer only communicates through the EtherCAT protocol, thus any sensors not using this needed to be converted. This was done by purchasing several terminal blocks/modules. The modules were purchased with the aim of being able to take in boolean values for initialisations and kill buttons, the external hip encoder, the two force sensors and the RS485 signal from the motion capture PC.

---

[1] Acknowledgement of Arnold Petorius, a PhD candidate in the Mechatronics Research Lab, who developed the visual tracking system.

The motor controllers (Ingenius Jupiter Drives) were pre-bought and have Ethernet ports to connect into the EtherCAT daisy chain. The encoders for the upperlinks on the robot were fed directly into these controllers.

A testing phase was done whereby all the sensors were ensured to operate accordingly and all data was arriving at the target computer without any errors.

3.7 Performance Testing and Verification

With all the infrastructure, embedded system, sensors and mechanical design ready, the robotic leg was powered on and systematically started. This was to ensure all systems were working as expected while reducing the risk of damage to the robot. During the initial testing and tuning of the motors, the gear coupling was removed so the motor could not drive the output.

This test was followed by a simple virtual compliance test, slowly incrementing the allowable torque output. Once the robot and all the systems were verified to behave consistently and robustly, a single leg was fully assembled.

A final outcome from this project was to show that the robot was capable of performing rapid acceleration manoeuvres. However, since no such controller currently exists, an alternative metric was used, vertical agility (in m/s) [2]. Vertical agility was the maximum height the robot can jump divided by the frequency at which it can hop. High leaping at high frequencies is considered to be more agile. This experiment was performed with both the monopod and biped configurations. Numerous robots and animals have had their vertical agility determined, thus the results gathered were compared against them to judge the platforms performance.

The proprioceptive accuracy of the leg linkage was determined by dropping the robot with a virtual compliance model onto a force plate. The actual and desired torques were compared to find the errors. It was assumed that losses were due to leg impedance.

To ensure the platform was suitably robust, the leg was put through several continuous hopping experiments. Once complete, the robot was disassembled to be visually inspected for any material yielding. Other properties of the platform were subjective and discussed, given the final design of the robot. From the above, the degree to which the platform meets the requirements set out in 3.1 were determined.

Chapter 4

Leg Design Realisation

This chapter covers the processes used to arrive at the transmission choice, the leg topology selected and the over all size of the platform. Furthermore, rapid acceleration motions are investigated through trajectory optimisation and the results are discussed herein. This chapter is mainly based on the conference paper published by the author [17] with the addition of several investigations.

4.1   Selection of Transmission and Compliance Scheme

As seen in the literature review, there are numerous actuation schemes to be chosen from. As noted in Chapter 3, the project is constrained to use pre-bought BLDC motors.

To compare the different actuation strategies, a web diagram is created in Figure 4.1 with the most important characteristics for this design. It should be noted that the following discussion on transmission selection works with all the information from Section 2.3.

Hydraulics and pneumatics are disregarded due to the expensive infrastructure. Furthermore, the requirement to use existing motors eliminates these as options for this project.

Since there is little research concerning the rapid acceleration manoeuvres, it is difficult to design the robot around any specific motion. It is desirable, for both modelling the SLIP template and for robustness, to have compliance in the system. However the transmission chosen should not constrain the type of motion the robot can perform. As mentioned in Section 2.3.2, the stiffness for SEA's is determined by the desired motion of a robot and thus is not useful for this project. It is thus decided that that virtual compliance is needed, such that the motion is not biased by a passive component and elasticity can be modulated as required.

To implement virtual compliance, the selected actuator must have high control bandwidth. The

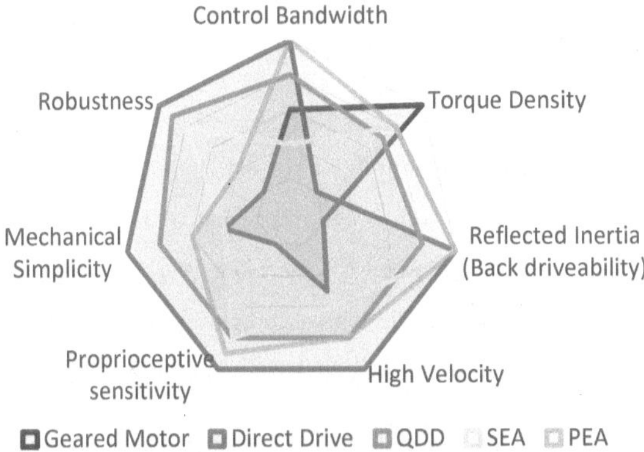

Figure 4.1: Actuation comparison summary. The further out on the web diagram indicates better performance. It can be clearly seen that the DD only suffers severly with the torque density. Oppositely the GM only performs well with torque density. The QDD is a good balance between the two. SEA's are complicated and negatively affect the control bandwidth. PEA's cannot assist with impact mitigation.

geared motor is almost immediately disregarded as it poorly performs in all fields except torque density. This leaves the option of either QDD or DD with possibly the use of a PEA.

PEAs do not affect the control bandwidth of the motors, accommodating for virtual compliance. Torque generated at the output is additive between the motor and elastic element. However, similarly to the SEA, it is unknown whether the use of a passive element will be beneficial or detrimental given that the motion for the platform has not be thoroughly investigated. It is decided that a passive element should be completely avoided and rather considered for use in a future iteration of the platform when there is a better understanding of rapid acceleration motions.

Direct Drive has high robustness, is mechanically simple and also has high control bandwidth. It also has no gears, therefore drastically improving the robustness. Furthermore, due to minimal losses, the proprioceptive sensitivity is also maximised. However, this comes at a major compromise to torque density. The torque peak for rapid acceleration motion is uncertain, thus using DD risks a lack of torque during the testing phase.

QDD on the other hand has been shown to forgo some of the advantages of DD to improve torque density. In experiments performed by Kalouche between DD and QDD (with N=7), it is seen that the proprioceptive sensitivity for DD is around 0.15Nm (a 12% error) while direct drive achieved a sensitivity of 0.4Nm (a 5.4% error) [19]. The percentage error for the force output is surprisingly less for the QDD. This is attributed to the several fold increase in output torque due to the single stage planetary gear.

The bandwidth of QDD is reduced from 70Hz for the DD, to 40Hz. This is detrimental for QDD, however, this control bandwidth is twice as great of a geared motor with 20Hz[19]. Lastly, the reflected inertia is increased by N^2, therefore QDD increases the inertia by a maximum of 100. This is acceptable considering any normal geared motor has a gear ratio of 50 and above which results in a reflected inertia of 2500 times larger.

Given the downfalls of using a single planetary gearbox, it is seen that the detrimental affects are still small compared to the use of a gear motor. Furthermore, the several fold increase in torque is highly desirable while ultimately improving the percentage error for the proprioception sensitivity. Thus, the QDD transmission is chosen with no use of PEAs or SEAs.

4.2 Foot selection

For this project it is proposed to introduce a foot that is not actuated as an actively controlled foot would add weight and mechanical complexity to leg mechanism reducing robustness. The foot for this project is chosen to be an extremely high damper to avoid the foot rebounding off the ground and dampen the initial impact. It is noted that the food should be easily removed and replaced as needed because the platform is to be used by numerous generations of students where various feet may be needed for different applications.

4.3 Leg Topology Identification

From the literature discussed, it is seen that there are many different leg configurations with their own benefits that assist in maximising the requirements in Chapter 3. The first process in developing the mechanical platform is to identify what linkage topology is optimal for a legged platform. This kinematic linkage will govern the overall dynamics and motion of the biped platform.

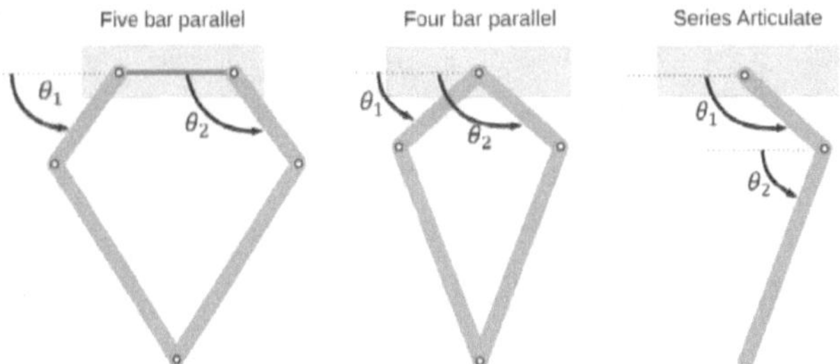

Figure 4.2: All three topology variations have two degrees of freedom. The four bar scissor linkage has seen use in ATRIAS [1] and Ghost Minitaur [21] while the series articulate is used in the MIT Cheetah [12].

As seen in the literature review (see the start of Chapter 2), the most common linkages used are the series articulate linkage, the scissor linkage, the prismatic linkage and the GOAT leg. The prismatic morphology is disregarded due to the pneumatic/hydraulic requirements. The overarching research group's focus is on rapid acceleration manoeuvres in two dimensions, thus the GOAT leg linkage is also disregarded.

A third variation of the scissor linkage mechanism is added where the motors are located adjacently instead of co-axially [50]. The four bar scissor linkage requires that both motors are coaxial, which drastically increases the width of the robot body. This would make the platform highly susceptible to moments in the roll axis which is undesirable for a biped. Furthermore, this project is limited to motion only in the sagittal plane, further emphasising a desire to avoid these moments and keep the hip width to a minimum. The modified scissor linkage mechanism is included in the up coming comparisons called the five bar scissor linkage mechanism. These three topologies can be seen in Figure 4.2. The series articulate is compared in this study as it has seen use in many of the inspired platforms. The optimal linkage is aimed to be identified through a derivation of maths, rather than mimicking what appears in nature.

The most important properties affected by the different topologies are:

- the functional workspace;
- the force transparency;
- the force amplification;
- the physical robustness, and;
- the load sharing between actuators.

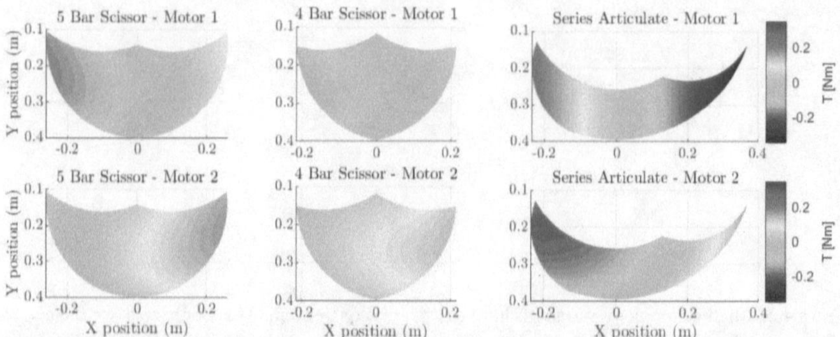

Figure 4.3: The load distribution characteristics of each linkage are shown. The 4 bar and 5 bar scissor linkages are seen to have a good load distribution for a 1N verticle force at the foot. However, the series articulate can be seen to have higher requirements from each motor, overall requiring higher rated torque motor to generate the 1N force. This figure also roughly depicts the shape of each linkages workspace.

4.3.1   Load distribution comparison

An additional metric investigated is how the load is distributed between the motors. This metric is beneficial, as by sharing the load equally across the workspace between two motors, a higher force output can be achieved. There are no configurations that demand the majority of work from one motor will another is barely used.

This is investigated by demanding a vertical unit force at the foot and the required torques are then plotted at all the different configurations of the workspace. To calculate the required motor torques, the linkage lengths are given an estimated leg length of 0.4 metres and ratio of 0.5. The resulting figures are shown in Figure 4.3.

As expected both the four and five bar scissor linkages distributed the load in a similar manner. However, the series articulate was far from optimal. The scissor linkages, for the most part, divided the load equally between motors. Meanwhile, the series articulate linkage has regions which placed a high load on one motor while the other is barely used. The affect of this is emphasised by the higher torque requirements for the series articulate linkage motors.

4.3.2 Singular Value Evaluation

The singular values provide insight into the scaling of vectors from one space to another, such as from the end effector forces at the foot to the motor torques at the hip. Interpreting the singular values of the Jacobian only concerns the geometry of the given linkage mechanism. This method is described in Appendix A, Section A.3 and has been used by Keneally [21].

The leg topology chosen needs to have a suitable trade off between:

- High proprioception, favourable as forces at the foot are more visible to the motor;
- High force production, desirable for torques to produce high forces at the end effector, and a;
- Large workspace - more space for the foot to operate in.

Good proprioception allows forces at the foot to be measured at the hip. This is important for impedance control and also enables the sensing of ground contacts. The transformation from motor torques to foot force should have a high gain but it is a trade-off with the proprioceptive property and a suitable balance should be identified.

The singular values can assist in the design and selection of these criteria in two ways. The **minimum singular value** at any location in the workspace indicates the visibility of forces at the foot to torques at the motor. A small minimum singular value indicates that forces at the foot are barely seen in the joint space [21]. To ensure accurate control and promote proprioception, it is desirable to maximise the minimum singular value. Oppositely, the **maximum singular value** represents the worst case amplification of the desired force at the foot to the required motor torque. This is ideally minimised to reduce the demands on the motors due to the leg topology. Thus, the maximum singular value should be minimised. It is easiest to find the trade off between the aforementioned values visually through graphs.

However, there is a different set of singular values at every configuration within the workspace. Furthermore, the ratio of the linkages also affects the singular values. Thus, it is required to look at how these values change due to the configuration and linkage ratio. This will also enable the selection of a ratio for the preferred linkage.

The singular values for the linkages are only identified for the central vertical workspace as bipeds exert most of the force in this configuration during leaping. Furthermore, this allows the affect of the different ratios to be identified simultaneously.

A comparison between the three aforementioned leg topologies can be seen in Figure 4.4, where an estimate nominal leg length ($L_1 + L_2$) of 0.5 metres is chosen to identify the morphology and linkage ratio.

It can be seen that the best ratio for the five bar and four bar scissor linkages are the same, roughly 0.35. While the best ratio for the Series Articulate is roughly 0.5. The two scissor linkages perform drastically better than the series articulate which has extremely high maximum singular values representing poor force amplification from the motors. The five bar scissor linkage has the lowest values across the entire workspace and linkage ratios for the minimum singular value out performing the four bar. The minimum singular values for the two scissor linkages are near identical and only affect the selection of the linkage ratios. Thus the figure clearly indicates that the series articulate is not desirable and the five bar is preferred.

4.3.3 Workspace Comparison

The workspace is also plotted for each different linkage mechanism to assist in the selection of the most suitable linkage ratio and leg topology. This comparison can be seen at the bottom of Figure 4.4. An example of the workspace for each linkage can be seen in Figure 4.3.

The standard shape of the workspace for the five bar scissor linkage can be seen in Figure 4.5. The area is calculated using a simple numerical approach. By dividing up the area the foot can move through into small rectangles and summing these slivers, the total area of the foot can be found.

Due to the limited knowledge on rapid acceleration manoeuvres, it is difficult to justify if an asymmetric workspace generated by the Series Articulate robot would be detrimental or not. This is a major downfall for the series articulate leg topology. Additionally, it suffers from high leg inertia due to a requirement for a motor at the knee. This can be avoided by using a four bar linkage as with the MIT Cheetah [27], however, this increases the complexity of the mechanical design and would require the motors to be co-axial. This creates a significantly wide hip for a biped which is further undesirable. Similarly, the four bar linkage creates a wide hip but does

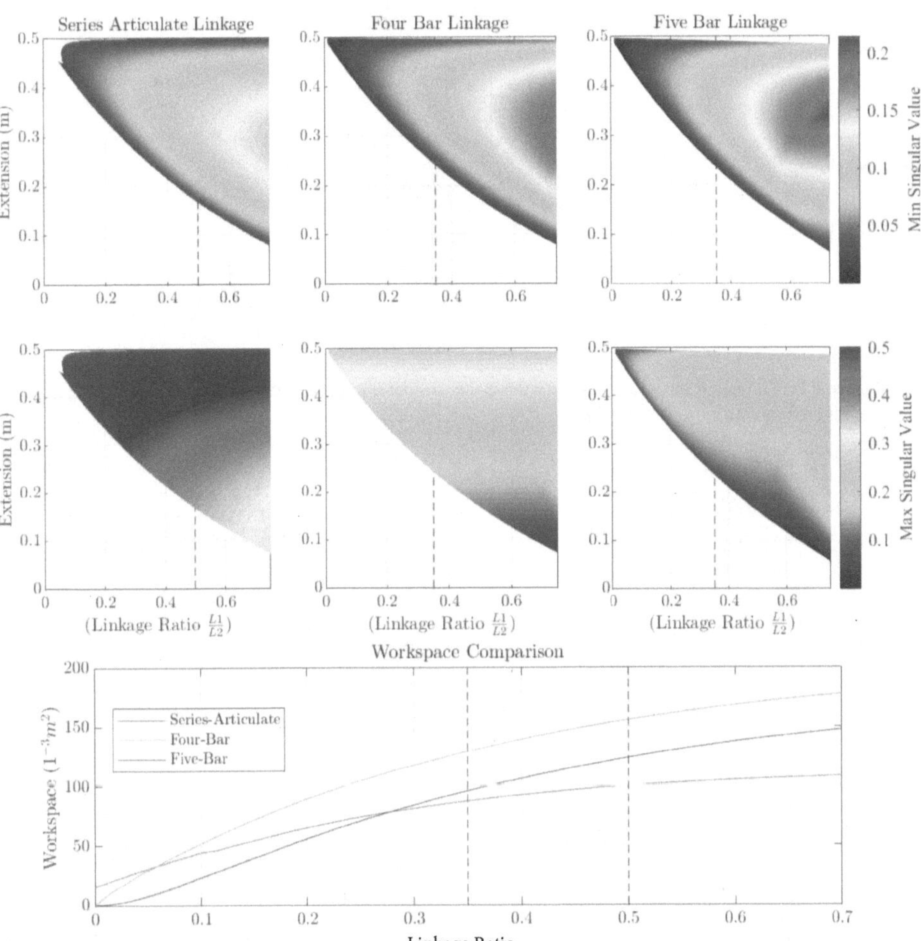

Figure 4.4: The singular values of each linkage are shown. The minimum singular value should be maximised to make the end affector forces more visible to the motors. The maximum singlar value should be minimised to improve the torque requirements at the motor to generate a force at the foot. The workspace is also shown with a changing linkage ratio (the shape of each workspace can be seen in Figure 4.3).

allow for a much larger workspace than the four bar linkage.

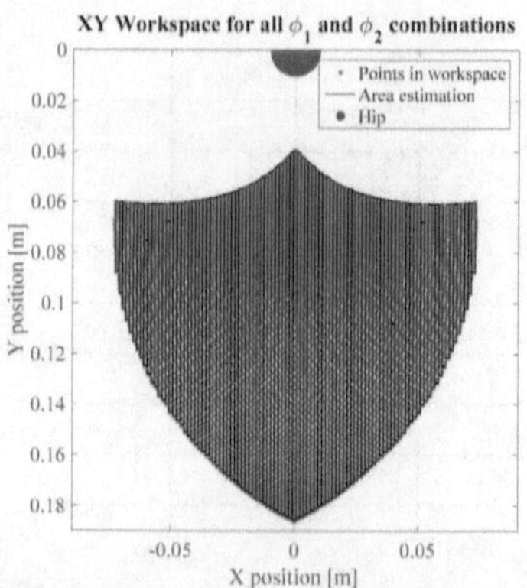

Figure 4.5: The five bar scissor linkage workspace. To determine the workspace area, the workspace was divided up into narrow vertical slices whose area's were added up.

Table 4.1: Leg topology Selection Table

Description	Max	5-bar	4-bar	SA
Workspace area	5	3	5	2
Physical Robustness	5	4	4	3
Force transparency	5	5	5	3
Torque magnification	5	5	4	1
Mechanical Simplicity	5	5	2	3
Total	25	21	20	17

4.3.4 Leg Morphology Selection

A selection criteria table for the morphologies can be seen in Table 4.1 and is generated from interpreting the results above. From the table, it is clear that the 5-bar and 4-bar are far superior to the series articulate linkage. The mechanical complexity of the 4 bar linkage was the deciding factor. This results from requiring motors and gearboxes to be located co-axial. In addition, as mentioned above, this creates an extremely wide hip. From the motors and gearboxes available, the author determined that the hip could be up to 700mm wide. Thus, the 5 bar scissor linkage morphology is chosen which avoids this issue.

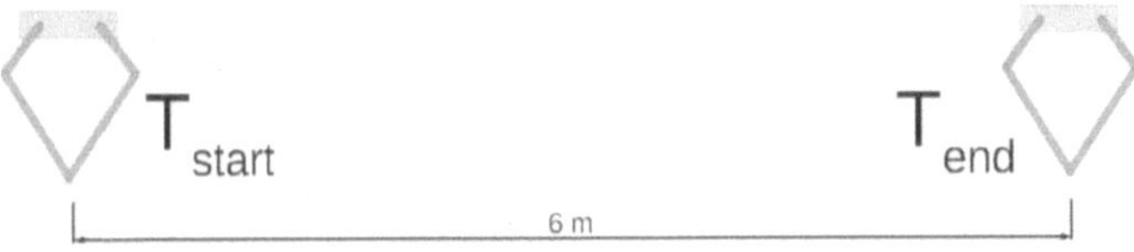

Figure 4.6: The *Tax Day* scenario where the robot is required to start and end in rest, but cover a distance in the shortest amount of time. This force the model to undergo rapid acceleration and braking motions.

4.4 Trajectory Optimisation

As seen above, the ideal leg length and ratio was determined. However, the next difficult trade off in the design process was between stride length and stride frequency (Spatial-temportal gait characteristics). To achieve high velocities large animals tend to have large stride lengths and low stride frequencies and oppositely for small animals. With short legs, a high mass specific torque is achievable but it compromises the leg length and reduces the duration the force can be output during each stride.

This was a difficult problem to solve as the physical platform design is inherently linked to the desired robotic motion [13]. The difficulty becomes finding the spatio-temporal gait characteristics to perform the desired task [57]. In the past, during platform design, only simple calculations were performed to identify the physical parameters for a task such as a simple hop. Furthermore, with complex manoeuvres it becomes impossible to identify thorough basic calculations or simulations. With the improvements in computational power, optimal control has now been used to enhance many different aspects of legged motion, as seen in Chapter 2.7.

The aim of this section was to investigate rapid acceleration manoeuvres and use trajectory optimisation in a novel way to assist in the selection of the nominal linkage length ($L_n = L_1 + L_2$) and gear ratio (N_g) given actuator limitations of pre-purchased motors (see Figure 4.8). The motors are the most important aspect of the robotic motion and thus these two other parameters need to be optimised around the motors.

The trajectory optimisation problem that was formulated is called the *Tax Day* scenario specified by Hubicki [15]. It is defined as minimising the time of a sprint, starting and ending in rest as depicted in Figure 4.6. Note that details for this section are expanded upon in detail in Appendix B.

4.4.1 Formulation

Numerical optimisation was used to find motions that are time optimal (minimise time) for acceleration and braking manoeuvres of bipeds. It was seen that the initial configuration and velocity of a model about to perform a braking manoeuvre has a major impact on how the model decelerates [16]. Thus, rather than investigating just rapid acceleration or braking manoeuvres independently, the entire motion was optimised in one problem. This was a minimum time sprint dubbed *Tax Day* by Hubicki [15] as mentioned above. This is the optimisation problem formulated.

4.4.2 Actuator and gearbox limitations

This project was limited to a motor that had already been purchased before the project was started. Thus, these motors are a constraint for this project. They are U12 T-motors and the relevant parameters can be seen in Table 4.2.

The gear boxes available are from *Matex* who supply single stage planetary gears ranging from 3 to 7. These limitations are taken into account in this chapter of the project.

4.4.3 Leg template Derivation

The five bar closed loop linkage that was chosen in Section 4.4 suffers from complex geometry. This transformation from the inertial frame to the joint angle frame can be seen in Appendix A. When the equations of motion (EOM) are derived using the Euler-Lagrange approach, the EOM grew to an enormousness size (the Coriolis matrix function file was 5mb). When this model was initially used in the optimisation problem (which is described below), each iteration of the solver took over 2 seconds which was completely impractical.

Previous studies have ignored the geometric complexity associated with the five bar linkage by ignoring L_{torso} (see Figure 4.8) when modelling their system [50]. This length was significantly large compared to the expected leg lengths and may lead to substantial errors. Hence, this simplifying step was ignored and a unique solution identified. Rather than modelling the system as a closed kinematic chain, each leg was split and modelled as two individual legs (see Figure

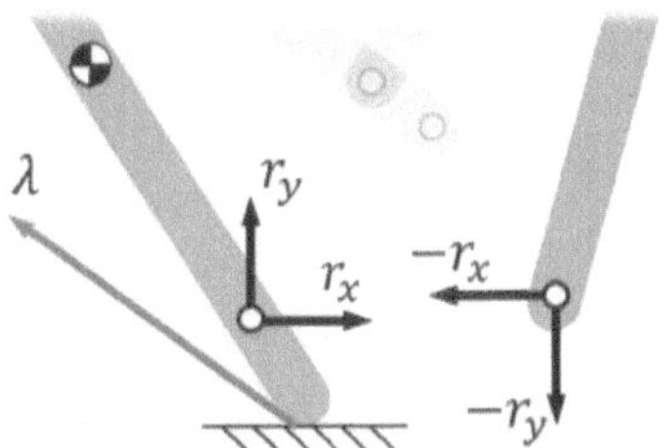

Figure 4.7: Forces r_x and r_y are solved by the optimiser to ensure that the to links remain together and are equal in magnitude but opposite in direction. The generated equations of motion are significantly smaller when modelling the system with the links unconstrained.

4.7). An external force acts at the desired connection point on each leg and are constrained to be equal but opposite. The optimiser then solves for the forces such that the leg connection remains together.

The weight of L_{Torso} was calculated by estimating the masses of components to be used: the motors, actuators, motor drivers and physical structure. To model the expected increase in mass of each link for the different linkage lengths, the mas was made linearly proportional to the length. The density and cross-sectional area were chosen to match a standard cylindrical aluminium extrusion. To increase the detail of the model, reflected inertia was also included [74]. The torso and all links are assumed to be rigid bars. These model parameters can be seen in Table 4.2.

The equations of motion were derived using the Euler-Lagrange method and presented in the commonly used Manipulator equation. The details of the derivation of the EOM for this system can be seen in Appendix A.

$$\mathbf{M(q)\ddot{q}+C(q,\dot{q})\dot{q}+G(q) = Bu+J_g}^T\lambda+\mathbf{J_c}^T\mathbf{r} \tag{4.1}$$

Where $\mathbf{q}$ is the generalised coordinates (see (4.2)), $\mathbf{u}$ is the control variables (see (4.3)), λ is the ground reaction force and $\mathbf{r}$ is the connection force with $\mathbf{J_g}^T$ and $\mathbf{J_c}^T$ as the Jacobian map into the generalised coordinate space for the connection force and GRF respectively. Figure 4.8 shows the generalised coordinates for the base model and in total the eleven degrees of freedom

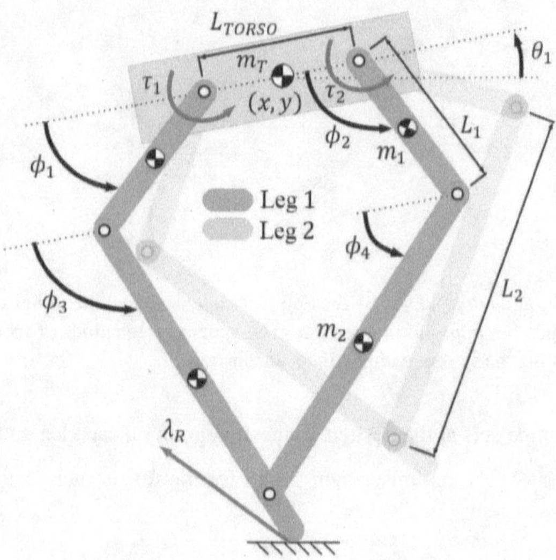

Figure 4.8: Forces r_x and r_y are solved by the optimiser to ensure that the to links remain together and are equal in magnitude but opposite in direction. The generated equations of motion are significantly smaller when modelling the system with the links unconstrained.

of the system can be seen below:

$$\mathbf{q} = [x, y, \theta, \phi_{11}, \phi_{12}, \phi_{13}, \phi_{14}, \phi_{21}, \phi_{22}, \phi_{23}, \phi_{24}]^T \tag{4.2}$$

$$\mathbf{u} = [\tau_{11}, \tau_{12}, \tau_{21}, \tau_{22}] \tag{4.3}$$

4.4.4 Discretization and Nodes

The trajectory optimisation problem was formulated by using the direct collocation technique and the trajectory was discretized into N nodes. Direct collocation methods is where the trajec-

Table 4.2: Model Details

Symbol	Value	Description
m_{torso}	6 kg	Estimate on required hardware
L_{torso}	0.09 m	Distance between links on torso
ρ_{link}	$2700\ \frac{kg}{m^3}$	Link density (Aluminium 6063)
A_{link}	$2.67 \times 10^{-4}\ m^2$	Cross-sectional area of links
μ	1	Ground friction coefficient
τ_m^{max}	7 Nm	Maximum motor torque output
ω_m^{max}	$450\ \frac{rad}{s}$	Maximum motor angular velocity

tory is broken into numerous nodes, where at each node all the constraints of the problem must be met [66]. In order to allow the optimiser to minimise time, the node length (time increment h) was allowed to vary.

$$0.1 h_{global} \leq h_n \leq h_{global} \tag{4.4}$$

Backward-Euler is an inaccurate scheme as it is implicit and tends to diverge significantly with large time steps. To try minimise this issue while keeping this problem to a realistic size, h_{global} was selected as 0.01 seconds (100Hz).

With this, the equations of motion were discretized using Implicit-Euler integration and took the form:

$$\mathbf{q}_n = \mathbf{q}_{n-1} + h_n \dot{\mathbf{q}}_n \tag{4.5}$$

$$\dot{\mathbf{q}}_n = \dot{\mathbf{q}}_{n-1} + h_n \ddot{\mathbf{q}}_n \tag{4.6}$$

By iteratively testing the problem, it was determined that by using 200 nodes, feasible solutions were found. The time the robot could run with 200 nodes varied between 0.2 and 2 seconds to cover the sprint distance.

4.4.5   Bounds and Constraints

In many design settings, actuators are the first elements that are chosen as they shape the overall behaviour of a robotic platform. Other physical parameters are generally chosen around the actuators limitations. Thus, as mentioned in Section 4.4.2, the actuators are limited to U12 T-motors[2] that have a high torque density and are readily available (see Table 4.2). A simple speed/torque motor model was included in the model [75].

An angular acceleration limit was added as it was discovered that the optimiser can take advantage of the integration error that arises from the Implicit-Euler scheme used. Furthermore, position limits for all the links were added and can be seen in Table 4.3.

In order to ensure that all variations of the model face the same task, constraints were placed on the starting and end conditions for the robotic platform. As per the *Tax Day* scenario specified by Hubicki [15], the model is constrained to start and end in the same configuration with

---

[2]Details can be found at: http://store-en.tmotor.com/goods.php?id=330

$\mathbf{q}(t_{start}) = \mathbf{q}(t_{end})$ (except for x) and zero velocity with $\dot{\mathbf{q}}(t_{start}) = \dot{\mathbf{q}}(t_{end}) = 0$. The sprint distance was chosen as six metres to allow the model to reach near steady state behaviour before braking. Furthermore, the laboratory in which the platform will ultimately be tested in is limited to this distance.

A minimum energy optimisation problem was set up over a single periodic step for the model. This was to compare how the optimal gear ratio and nominal leg length differs from an optimal minimum time sprint. The model was defined to start with zero y velocity representing the apex of the gait [69].For the minimum energy problem all links were specified to have zero angular velocity at t_{start}. The configuration was set to have the biped legs furthest apart at t_{start} and to end with the legs in the opposite position (i.e, $\mathbf{q}_{leg2}(t_{end}) = \mathbf{q}_{leg1}(t_{start})$ and vice versa).

4.4.6   Ground Contacts

The ground contacts are modelled as inelastic collisions with sliding according to the coulomb friction model. Ground contact forces are determined using a set of complementarity constraints proposed by Posa et al. [76]. However, Fletcher et al. found that the addition of slack variables improves the convergence for complementarity equations [77] and has been implemented successfully before [14]. The constraints seen from (4.7) to (4.18) must be held and are all expressed in the inertial frame (x and y). The slack variables are positive real numbers:

$$\alpha_{1,n}, \alpha_{2,n}, \beta_{1,n}, \beta_{2,n}, \kappa_{1,n}, \kappa_{2,n}, \gamma_{1,n}, \gamma_{2,n} \geq 0 \tag{4.7}$$

Equations (4.8)-(4.10) states that $\lambda_{y,n}$ (normal force) only acts when the foot is in contact with the ground where $\phi(\mathbf{q_n})$ is the contact point height and cannot penetrate the ground.

Table 4.3: Model Bounds

Gen. Coordinate	Bound	Description
$\ddot{\theta}, \ddot{\phi}_{11}, \ddot{\phi}_{12}, \ddot{\phi}_{13}, \ddot{\phi}_{14},$ $\ddot{\phi}_{21}, \ddot{\phi}_{22}, \ddot{\phi}_{23}, \ddot{\phi}_{24}$	$-5000 \rightarrow 5000 \frac{rad}{s^2}$	Acceleration limit for all rotation coordinates
ϕ_{11}, ϕ_{21}	$0 \rightarrow 90^o$	Link 1 angle bounds
ϕ_{12}, ϕ_{22}	$90 \rightarrow 180^o$	Link 2 angle bounds
$\phi_{13}, \phi_{14}, \phi_{23}, \phi_{24}$	$0 \rightarrow 180^o$	Link 3 and 4 angle bounds

$$\phi(\mathbf{q_n}) = \alpha_{1,n} \tag{4.8}$$

$$\lambda_{y,n} = \alpha_{2,n} \tag{4.9}$$

$$\alpha_{1,n}\alpha_{2,n} = 0 \tag{4.10}$$

$$\lambda_{x,n}^{+} = \beta_{1,n} \tag{4.11}$$

$$\gamma_{1,n} + \psi(\mathbf{q_n},\dot{\mathbf{q}}_n) = \beta_{2,n} \tag{4.12}$$

$$\beta_{1,n}\beta_{2,n} = 0 \tag{4.13}$$

$$\lambda_{x,n}^{-} = \kappa_{1,n} \tag{4.14}$$

$$\gamma_{1,n} - \psi(\mathbf{q_n},\dot{\mathbf{q}}_n) = \kappa_{2,n} \tag{4.15}$$

$$\kappa_{1,n}\kappa_{2,n} = 0 \tag{4.16}$$

$$\mu\lambda_{y,n} - \lambda_{x,n}^{+} - \lambda_{x,n}^{-} = \gamma_{2,n} \tag{4.17}$$

$$\gamma_{1,n}\gamma_{2,n} = 0 \tag{4.18}$$

Equations (4.11)-(4.16) ensure that the x GRF, $\lambda_{x,n}$ ($\lambda_{x,n} = \lambda_{x,n}^{+} - \lambda_{x,n}^{-}$), cannot act in the same direction as the foot velocity during slipping, where $\psi(\mathbf{q_k},\dot{\mathbf{q}}_k)$ is contact point velocity. Equation (4.17) and (4.18) keep the contact forces within the friction cone when there is no slipping, where μ is the coefficient of friction. To allow the optimiser to search a wider solution space, the complementarity equations were not directly constrained but instead penalised in the cost function. That is, (4.10), (4.13), (4.16) and (4.18) were added into the cost function (see Section 4.4.7) and must be reduced to zero for a solution to be considered feasible. Subscript n represents a discrete node.

These complementarity constraints allowed the optimiser to explore the entire solution space without the limitation of a predefined contact sequence. The above equations were used for both foot contacts on the biped, hence repeated twice.

4.4.7 Objective function

To optimise for a sprint (*Tax Day* scenario), the time taken for the task to be completed was located in the objective function and is defined as:

$$t_f = \sum_{n=1}^{N} h_n \tag{4.19}$$

By forcing the robot to start at the origin and end 6 metres away while minimising time, rapid acceleration and braking motions of the robot are generated by the optimiser.

Due to the complexity of the model it was not feasible to simultaneously solve motions for rapid acceleration manoeuvres and the optimal physical parameters (N_g, L_n). Thus, a brute force approach was used to find the optimal physical parameters (see Section 4.4.8).

As mentioned in Section 4.4.6, several complementarity equations are included in the objective function as a penalty (see (4.20)). To avoid affecting the minimum time t_f in the objective function, this penalty must be reduced to zero for the solution to be feasible. The objective function takes the form of (4.21). A scaling factor ρ_1 (with a value of 10 000) was used to ensure the optimiser minimised the complementarity penalty to zero.

$$\delta = \sum_{n=1}^{N} (\alpha_{1,n}\alpha_{2,n} + \beta_{1,n}\beta_{2,n} + \kappa_{1,n}\kappa_{2,n} + \gamma_{1,n}\gamma_{2,n}) \tag{4.20}$$

$$J = t_f + \rho_1 \delta \tag{4.21}$$

The objective function for steady state motion (see (4.23)) excludes the time cost and rather includes the torque squared value multiplied by the time duration at that node for each actuator. This is then divided by the total stride length (x_N). The torque squared approximation has been shown to be a reasonable criterion for systems with electrical motors [78]. The scaling factor ρ_2 is chosen to be 0.01 since the COT is significantly large and should be optimised only after the GRF penalty has been minimised to zero.

$$T = [\sum_{n=1}^{N} h_n(\tau_{11,n}^2 + \tau_{12,n}^2 + \tau_{21,n}^2 + \tau_{22,n}^2)]/x_N \tag{4.22}$$

$$J = \rho_2 T + \rho_1 \delta \tag{4.23}$$

4.4.8 Implementation

The system described in the above sections was implemented in the General Algebraic Modelling System (GAMS) which has shown promise in large systems such as those used in [16]. GAMS is significantly faster than the MATLAB environment as it has been purpose-built for handling large-scale non-linear optimisations. The problem was optimised by using two gradient decent solvers, first being IPOPT and then the results passed to CONOPT, exploiting the advantages of both solvers [79].

The equations of motion for the biped model was generated by using a Matlab script (described in Appendix A.2). An additional script was coded by the author to convert the Matlab function files to a format compatible with the GAMS environment.

The solver was initiated with a uniformly distributed random seed for the generalised coordinates as to not bias the solutions and ensure that the solution space was well explored. All other variables were initialised to a small non-zero value of 0.01 as this improves the convergence rate [15].

As mentioned in Section 4.4.7, a brute force method was used to identify the optimal nominal leg length and gear ratio for a six metre sprint. This involved running 100 seed points for the several different leg lengths and gear ratio combinations. The optimiser operated on the cost function over all generalised coordinates (and first derivatives), control inputs, ground reaction forces and node durations from node one to N (see (4.25)). In this section the number of nodes used was 200 which makes this problem significantly larger than the 40 used by Xi et al. and Posa et al. [69][76].

$$\underset{Q}{\text{minimize}} \ J(\mathbf{h}, \mathbf{q}, \dot{\mathbf{q}}, \mathbf{u}, \lambda, \mathbf{r}) \tag{4.24}$$

$$Q = \{h_1...h_N, \mathbf{q}_1...\mathbf{q}_N, \dot{\mathbf{q}}_1...\dot{\mathbf{q}}_N, \mathbf{u}_1...\mathbf{u}_N, \lambda_1...\lambda_N, \mathbf{r}_1...\mathbf{r}_N\} \tag{4.25}$$

Three planetary gear ratios that are readily available, such as those used in [19], were tested in the model (3, 5 and 7). Nominal leg lengths ranging from 200 mm to 800 mm were tested for each gear ratio. Each model variation was optimised from roughly 100 seeds. In total 1600 seeds were run with 315 feasible solutions for the tax day scenario. For the steady state tests,

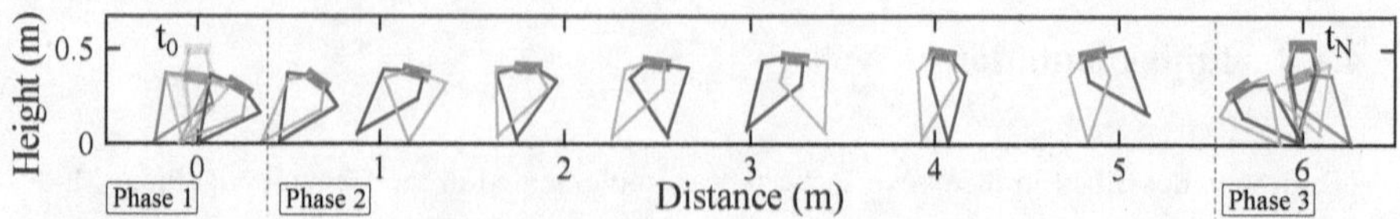

Figure 4.9: Optimal minimum time sprint model with $N_g = 5$ and $L_n = 0.5$m. The motion is broken into 3 different phases. Rapid acceleration occurs in Phase 1, constant acceleration in Phase 2 and rapid braking in Phase 3. The model swings the legs out of phase to cancel any torques around the torso.

1800 seeds were run with 824 solutions.

4.4.9 Optimisation Results

For the sake of brevity, the models with a gear ratio of 3 are referred to as *Model3*, 5 as *Model5* and 7 as *Model7*. It should also be noted that most of the comparisons made here are between the optimal solutions for each individual model variation. In addition to these results, there were 294 other feasible solutions but whose performances are substandard.

The sprint times for each tax day trajectory are compared in Figure 4.10. A global minimum for the sprint time of each gear ratio model was found: Model3 with $L_n = 0.4$ m; Model5 with $L_n = 0.5$ m; Model7 with $L_n = 0.6$ m. However, the optimal between the aforementioned models was Model5 and the full sprint motion can be seen in Figure 4.9. The motion of the various models can be seen in the video here.

At the extreme lengths, large linkages had more mass, decreasing the stride frequency and putting stain on the actuators, while short links had less mass and required higher stride frequencies. Further implications of this are clearly seen in the parabolic shape of the minimum time solutions in Figure 4.10, regardless of the gear ratio used.

All models completed the sprint within half a second of each other, with the optimal gear ratio models sprint times varied by only roughly 0.2 seconds. In order to investigate this further, the motion was broken into three different periods. Phase 1: rapid acceleration; Phase 2: constant acceleration; Phase 3: braking.

Rapid Acceleration Phase

It was assumed that model7 would excel in the rapid acceleration phase, seen in Figure 4.11, due to the model's high mass-specific force output. Nevertheless, all the models performed

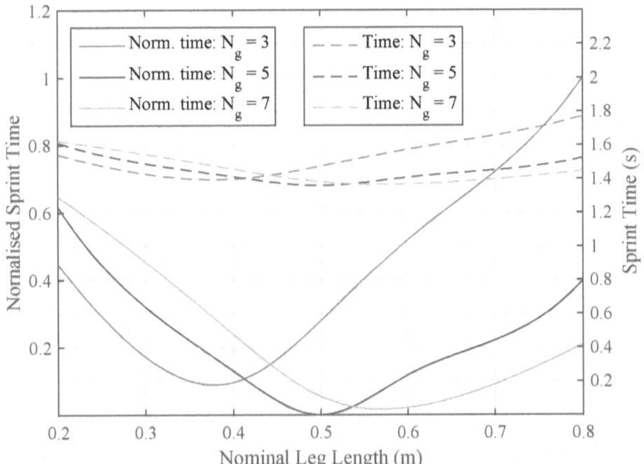

Figure 4.10: The normalised minimum time for six metre sprints are shown as a function of the nominal leg length (L_n) and gear ratio (N_g) used in different models. It can be seen that large gear ratios improve performance with longer links. However, the optimal configurations of each gear ratio and leg length varies by only roughly 0.5 seconds.

equally. The friction model had $\mu = 1$ which constrained $\lambda_x \leq \lambda_y$, limiting all models to a GRF vector at 45° or more. Model7 could easily exert an excessive normal force and perform large leaps, achieving high initial accelerations in the x direction. However, by performing such a manoeuvre, the model's feet would be out of reach of the ground for a longer period (in a ballistic trajectory). This negatively effects the x acceleration as foot contact was the only way to add energy into the system in that direction.

Figure 4.12 shows how the above mentioned scenario was avoided. All the models initially dropped their centre of mass low by taking a crouched position similar to that of a human at the start of a race (see Figure 4.9) and then kept their centre of mass the same height above the ground (normalised to the models L_n) [70]. Maintaining the COM height occurred not only in the rapid acceleration phase but also in the constant acceleration period, keeping the ground within the extension range of the leg. This enabled multiple ground contacts.

This suggests that the motors are over-powered and that a limit exists where more torque no longer affects the acceleration capabilities, contradicting the original speculations. There seems to be two interlinking factors that govern this bound. One factor is mentioned above that concerned the desire to keep the feet within reach of the ground. This puts a limit on the Y impulse that can be exerted on the models during each step, which in turn limits the X acceleration in

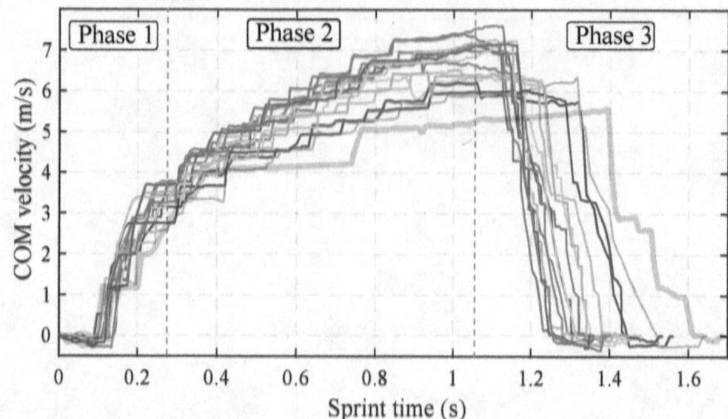

Figure 4.11: The centre of mass (COM) *x* velocity trajectories for all gear ratios and nominal leg lengths from 0.3 to 0.7 metres are plotted. $L_n = 0.2$ and $L_n = 0.8$ models are intentionally disregarded due to poor performance. The motion was broken into 3 different sections. A non-optimal solution is shown by the grey line which failed to keep the COM near the ground.

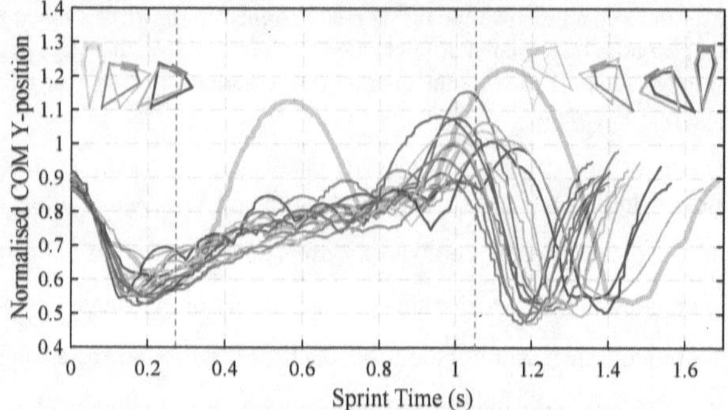

Figure 4.12: The COM height (*y* direction) was normalised with respect to the models nominal leg length ($y_{com,i}/L_{n,i}$). An optimal height for the COM exists that was common for all models during the rapid and constant acceleration phases. In the last stage, all the models lower their COM with several models performing small leaps to increase the frictional force for braking. A breakdown of the optimal platform motions for phase 1 and 3 are displayed. A non-optimal solution is shown by the grey line which failed to keep the COM near the ground and hence performed poorly.

that step. Since all the models accelerated at roughly the same rate in phase 1 (see Figure4.11), it indicates that the impulse bound was hit. The second factor was the friction coefficient. Figure 4.13 shows how the GRFs (for the optimal sprints of Model3, Model5 and Model7) lie on the friction cone in the rapid acceleration phase. This emphasises that the models are exerting the maximum possible GRF force in the X direction and should μ increase, more force could be utilised for X acceleration.

Model5 with $L_n = 0.5m$ seen in Figure 4.11 and Figure 4.12 (as a thick grey line) shows a non-optimal sprint, where the model performs large leaps rather than keeping it's COM near to

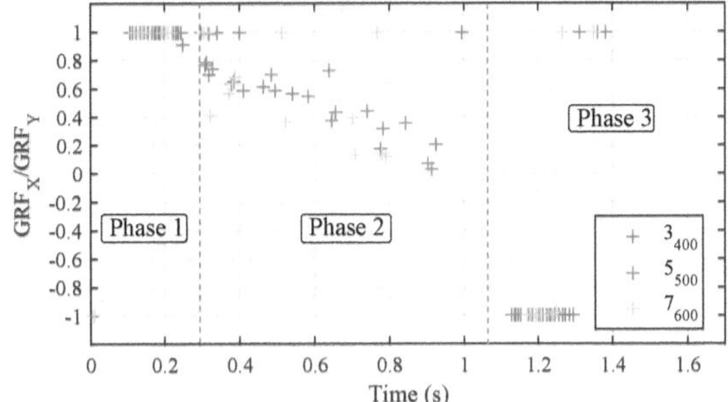

Figure 4.13: λ_x/λ_y for the optimal solutions for each gear ratio model is shown. During the rapid acceleration and braking phase the forces lie on the friction cone. The slow decrease in the GRF ratio indicates that the models are approaching steady state, however the sprint distance was too short. The small leap performed by the models can be seen on the border between Phase 1 and Phase 2.

the ground. This resulted in a significantly increased sprint time.

Braking Phase

The optimiser found a common solution for all models during the braking phase. That was, performing a small leap (clearly seen at the end of the constant acceleration phase in Figure 4.12), throwing the legs forwards. The legs made contact with the ground at 45° to reduce the moment around the COM caused by the GRF. The leap enabled the models to increase λ_y on landing and produce a high initial friction force. Furthermore, Figure 4.11 shows that all the optimal solutions de-accelerated equally. In this braking manoeuvre, all models lie on the friction cone and experience slipping, hence the GRF's are at 45°.

Constant Acceleration Phase

After the initial phase, a constant acceleration period occurs where, on each stride, the models increase their COM velocity. It was herein that the different variations in the optimal time are most effected, with a span of up to $2m/s$ variation in top speed. This showed that the rapid manoeuvres were not the deciding factor for time optimal sprints as expected. Figure 4.13 shows that the models approach steady state, seen by the slow decrease in λ_x/λ_y over time. When the GRF ratio was roughly equal to zero the model would be travelling at steady state with no net force in the x direction. However, the sprint distance was not sufficiently far, as a GRF ratio of zero was not reached.

Minimum Energy

To compare the results gathered from the minimum time sprint, the same models were opti-

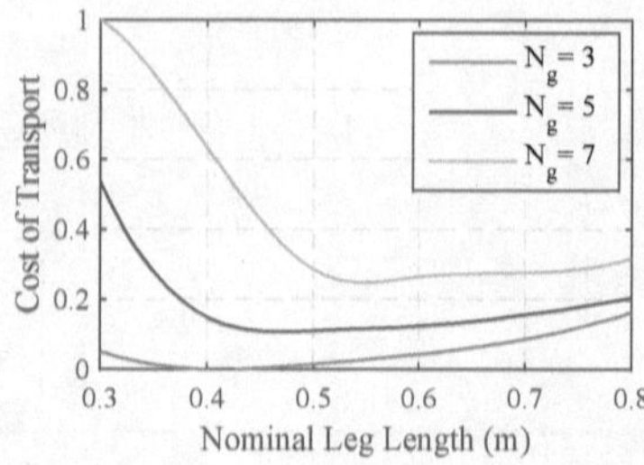

Figure 4.14: The COT is normalised with respect to the minimum and maximum values found. The optimal configuration is $N_g = 3, L_n = 0.4$ m. Interestingly the optimal leg lengths for each gear ratio model are similar to those discovered in the minimum time sprints and a similar parabolic curve occurs.

mised over a steady state, minimum energy problem. The steady state model was limited to half a stride and the optimal solutions revealed that the minimum energy models were very similar to those used for minimum time sprints. Figure 4.14 follows a similar trend to Figure 4.10, where a local minima for each geared model exists and are located in a similar position to those for minimum time sprints. This hints at the possibility that parameters defined for steady state platforms may be similar to those for constant acceleration manoeuvres.

4.4.10 Optimisation Conclusion

In this section, we aimed to identify the optimal nominal leg length and gear ratio for use in a minimum time sprint of a biped, given the limitations of a pre-specified actuator. The scope of this section limited the model to the sagittal plane to simplify the optimisation problem. A realistic and common linkage morphology was used to improve the realism of the model as it will ultimately provide insight into the mechanical design of a biped platform. Due to the complexity of the optimisation problem, a large number of seeds were required to generate the results for *Tax Day* scenario, with a convergence rate of 19.32%, where non-convergence indicates that the constraints of the problem could not be satisfied. This is considered a good convergence rate as other researches performing large-scale optimisation have only achieved a convergence rate of 10% [80].

The robotic platform performed manoeuvres similar to those used by the bipeds biological counterparts, such as crouching before rapid acceleration and throwing it's legs forward during rapid braking. The entire motion for the optimal model, $L_n = 0.5m$ and $N_g = 5$, can be seen in

Figure 4.9 along with these aforementioned manoeuvres.

Given the optimal solutions for each gear ratio model, the author's initial prediction that models with a high mass-specific force would perform better was disproved. The time variation was only roughly 0.02 seconds. We found that a bound exists where the actuator power no longer governs the rapid acceleration performance. Rather, keeping the body low to the ground for multiple foot contacts and the friction coefficient (μ) became the limiting factors. The braking phase was governed by the coefficient of friction where force output does not play a significant role. Robotic engineers designing for rapid acceleration should focus more on foot contact friction and not solely on mass-specific force output.

Ultimately, the author could not break down all the fundamental elements at play due to the extreme complexity of acceleration manoeuvres and how the physical characteristics factor into the overall dynamics of the robot. Rather it was demonstrated that use of large scale optimisation could investigate rapid manoeuvres that are currently not understood. The author was further able to show the optimisation techniques could be used to gain valuable insight into parameters required for the physical biped model.

4.5 Summary

This chapter aimed to identify the actuation scheme, leg topology and general sizing for the robot. Through a thorough investigation, the most suitable actuation scheme, Quasi-Direct Drive, was selected. This improved the mass-specific force output as a small cost to actuation bandwidth and increased reflected inertia. Furthermore, a leg topology and linkage ratio was identified that distributed the foot forces equally between motors, while promoting proprioception for foot force sensing.

With this base, trajectory optimisation was used to investigate rapid acceleration motions of a biped model. This work provided the optimal nominal leg length of the linkage and the best gear ratio for rapid acceleration motions. It was found that the motors were overpowered and that should a higher acceleration motion be desired, the foot friction force should be increased. A paper concerning this section has been published in the 2018 International Conference on Robotics and Automation [17].

Chapter 5

Mechanical Design

Designing the mechanical platform was an extended process with numerous design iterations and reviews. Due to financial limitations and the turn around time of the fabrication workshop at the University of Cape Town, only a single platform could be made. This constraint restricted the author from working with adequate prototypes and raised the importance of ensuring the design was flawless before fabrication. This platform should not be seen as a final indefinite design of the robotic leg, but rather one step in several to achieve a robust, lightweight, high speed leg for the next decade of students to work with.

The design process is highly cyclic but the author will attempt to present the process as a linear story. The reader should note that the majority of the design decisions were excluded from this report and only the most notable design steps and changes will be shown in this section which led to final platform.

Lastly, a new constraint was placed at the last minute before fabrication. Previously, the U12 T-motors were the constraint as they had been purchased already. However, given a fellow PhD student's requirements for as much power output as possible (due to auxiliary components to carry), U13 T-motors had been purchased. The author was required to modify the final design to use the larger, more powerful motors. However, during experimentations, the negative characteristics of the U13s outweighed the torque output benefits. This is explained in Chapter 8 and the design was modified to use the U12s as originally planned. The reader should note that this chapter concerns the use of the U12 motors.

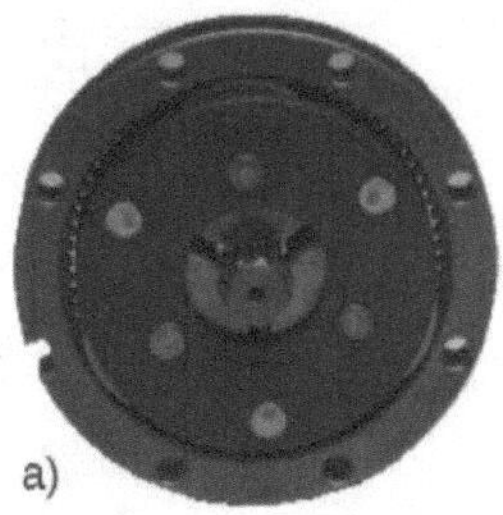
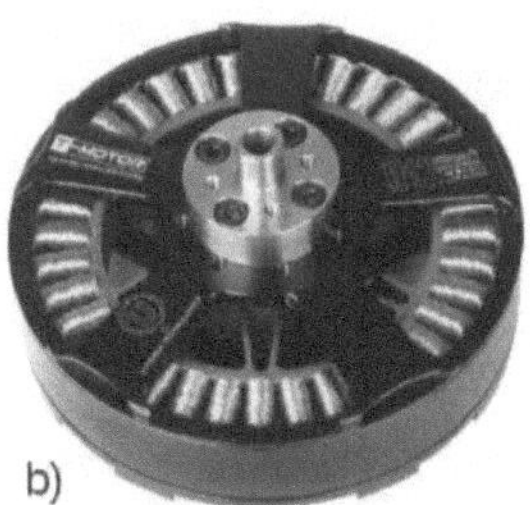

Figure 5.1: The following parts were provided for the project and are seen as constraints. a) Matex steel planetary gearbox (three versions with a reduction ratio of 3, 5 and 7). b) The T-motor U12 brushless DC motor with a rated torque of 5.3Nm

5.1 Requirements and specifications for Design

From the project requirements that were identified in Chapter 3, more precise requirements are detailed here concerning the physical platform design:

- Modular leg for repositioning.
- Minimise cost of manufacture (avoiding complicated parts to manufacture, expensive bearings, large metal plates).
- Minimum safety factor of 2.
- Design should be such that unavoidably expensive parts will not be damaged.
- Easy to assemble and exchange parts.
- Robust against impacts.
- Allow for strengthening of components should they be too weak.
- Space for embedded system components.

To reduce the cost of manufacturing, it is ideal to avoid parts that are difficult to manufacture as well as expensive materials. The main constraints for the design are those of having to use the U12 T-motors (U12 KV90 T-motors) and the planetary gearboxes provided by MATEX as seen in Figure 5.1. Throughout the numerous design iterations of the robot, the above requirements were accounted for.

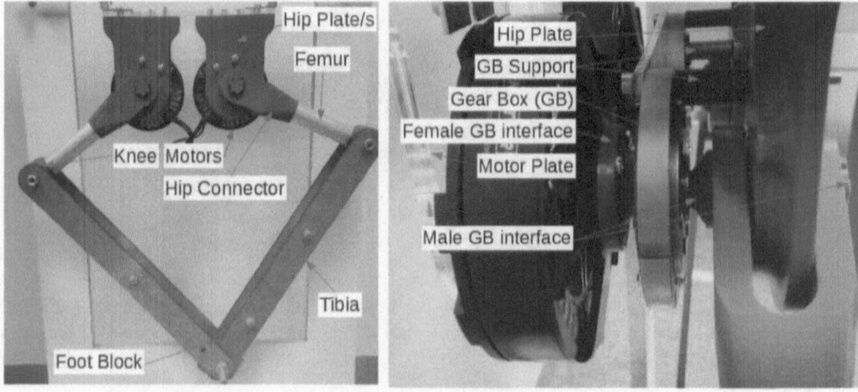

Figure 5.2: Images of the leg prototype with labels for the different parts.

Figure 5.3: Render of the prototype from SolidWorks.

5.2 Prototype

A preliminary design was drawn up and on the 6th iteration, a prototype was made to verify the design. The prototype tackled several key design areas. Those being:

- Gearbox interface;
- Linkage joints, and;
- Assembly issues.

The prototype was made from 3D printed parts, laser cut hardboard and aluminium tubing. The SolidWorks design and the actual assembly can be seen in Figure 5.2 and 5.3. Note that the names used in the figure will continue to be used throughout this design section.

From the prototype, a number of flaws were identified. The *Male GB interface* used to drive the output linkage, *Hip Connector*, was extremely complicated to manufacture with three different splines on a single part. In addition, the shaft was located by a single rolling element ball bearing held within the *Hip Plate*. The single bearing allowed significant play of the linkage out of the sagittal plane, indicating that a system with two or more bearings was needed to make the leg rigid. Furthermore, the *Hip Connector* would be difficult and expensive to machine, given the splines interfacing with the *Male GB interface* and the large slot to fit around the *Hip Plate*. This driven link needed to be completely redesigned. An explosion view of this part can be seen in Figure 5.4 a).

It was noted that significant bending took place between the *Motor Plate* and the *Hip Plate*. This was because there were no cross supports between the two plates. The gearbox was designed to float on the *GB Support* pins and takes no radial loads. Knowing that the leg would be under significant stress, it is evident that supports should be added to create a rigid hip structure.

The *Leg Connectors* did not achieve the rotational range desired by the author as they would impact the *Hip Plate*. The plate has to be redesigned to allow the *Leg connector* to rotate higher. The *GB Support* pins were designed as hollow steel tubes with a bolt going down the centre, however this did not provide accurate alignment and needed to be adjusted.

5.3 Gearbox Coupling Design

Several design iterations were cycled through to identify a suitable transfer mechanism into and from the gearbox. In design iteration 9, seen in Figure 5.4 (b), the mechanism was designed with only a single spline to interface to the gearbox and four bolts to locate onto the leg connector. Unfortunately, the manufacturing of the spline by a local gear cutting firm was excessive and would cost over R1500 a piece. Such an expensive part is impracticable and if made with slightly incorrect tolerances, may induce additional backlash. It was ultimately decided that the final design should make use of the couplings made by the gearbox manufacturers, *Matex*. In Figure 5.4 c), the purchased shaft can be seen with an interfacing metal sleeve and key. The shaft purchased was the only available option offered by *Matex* with a yield strength of

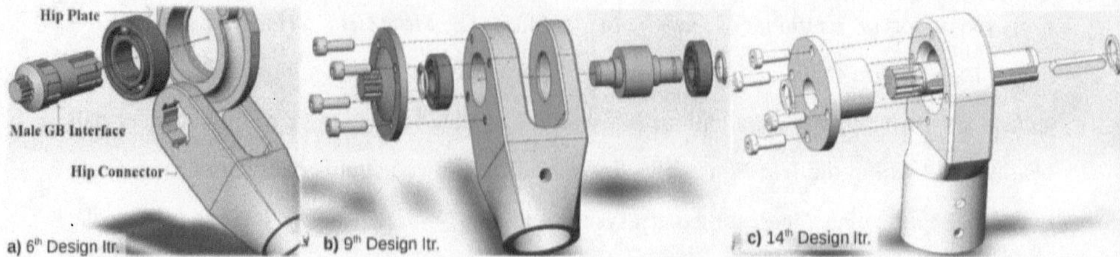

Figure 5.4: The various designs for the interface into the female side of the gearbox and the transmission design into the leg. **a)** The 6th design was constructed in the prototype but the single bearing allowed for large amounts of out of plane motion. **b)** Design 9 was ideal however, manufacturing the male spline at a gear cutting firm was outside of the budget for this project and may cause additional backlash. **c)** The 14th design is used in the actual platform whereby the male spline is purchased from the gearbox supplier.

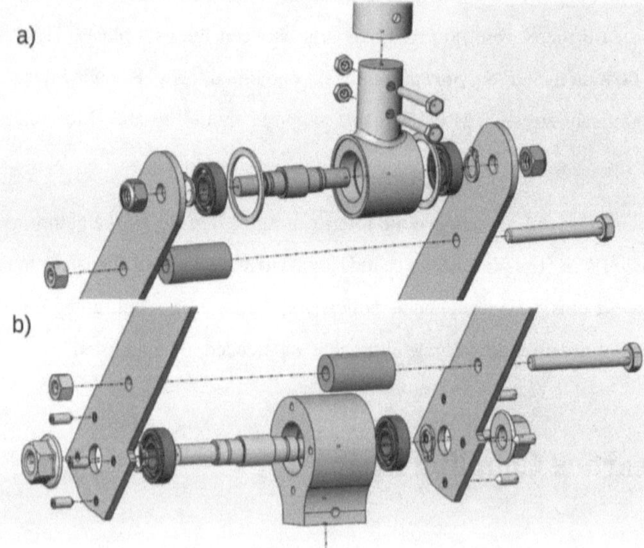

Figure 5.5: a) Exploded view of the knee assembly. **b)** Exploded view of the foot assembly

900MPa.

5.4   Knee and Foot Design

The knee and foot design remained mostly unchanged from the prototype design iteration. The joints with the double bearing were seen to have negligible friction and could withstand high out of plane forces with no visible deflection. The tibia plates were chosen to be made from easily cut metal sheets while the femur link was selected as a standard aluminium tube section.

The foot block was designed such that various foot connections could be attached if needed with a standard shape. For the purposes of this project, however, a rubber food is employed to provide adequate friction and damping upon landing which can be seen in Section 5.7, Figure 5.9b.

5.5 Parts for Safety

A major concern was that of the robot legs swinging into the hip. To ensure this did not happen, hard stops were created made from an aluminium block with a section of rubber for damping. Furthermore, a limit switch is located above the hard stop such that if the leg is about to hit the robot body, the motors would cut out (see Section 7.2.2 for their implementation). The hard stops and limit switches can be seen in the final CAD render.

5.6 Final Design

Additional items that had to be accounted for in the design were the location of the encoders, electronic speed controllers and shunt resisters for the motors. The final design with all these components can be seen in Figure 5.6.

5.7 Part Strength Analyses

There are several critical failure points that exist in the platform that were analysed to ensure sufficient safety factors for the robot. The main issues of concern were the output of the planetary gear box and the upper links of the legs due to the high bending moment that would be present. The reader should note that the strength analyses and subsequent changes to the design was a part of an iterative process. Changes were made when parts were too weak and vice versa.

Figure 5.6: Left: The final render of Baleka with the ESC's and encoders mounted. **Right:** A single modular leg that would be initially assembled.

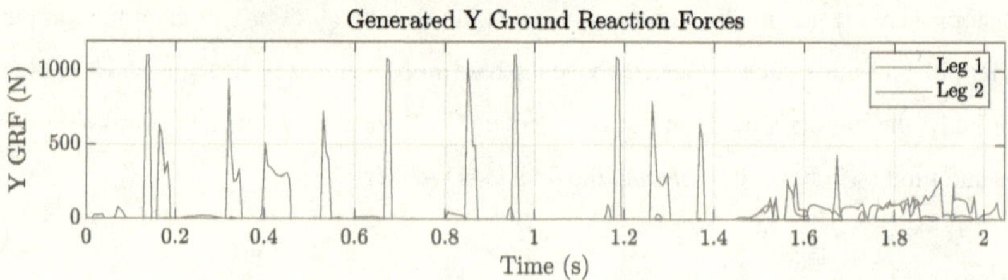

Figure 5.7: The ground reaction forces of the bipedal model generated from the trajectory optimisation problem. The force peaks near 1100N on impact. This problem was solved with 400 nodes to improve the GRF resolution.

Using the optimisation results generated in Section 4.4, the internal forces between the links were calculated during the rapid acceleration manoeuvres. The trajectory generation optimisations yields all the generalised co-ordinates (up to the second order) and all external forces. By using these results and deriving a completely unconstrained model (with identical properties), all the internal forces linking the unconstrained bodies together could be solved. The highest forces seen where those when the foot impacted the ground, reaching a maximum of 1100N. The ground reaction forces generated from the trajectory optimisation can be seen in Figure 5.7.

These calculated forces were used in SolidWork's Finite Element Analyses (FEA) to calculate the expected stress the platform will endure. Some of the critical components of concern were: the output shaft of the gearbox and upper leg that undergoes significant bending moments; the support plates that must support the robot; lower leg plates that may buckle and the foot block that must withstand high impacts.

To the best of the author's ability, the materials available in SolidWorks were selected that most closely match the physical properties of the actual material used. The results of some of the critical parts can be seen in the below figures.

The factor of safety (FoS) is viewed when analysing parts. It is defined as:

$$FoS = \frac{\sigma_y}{\sigma_w} \tag{5.1}$$

where σ_y is the yield strength of the material and σ_w is the working stress as some point on the part. The property indicates clearly how much additional load a part can handle from the working load. To ensure a robust robot, the safety factor should not go below 2 (thus, two times the working load).

5.7.1 Gearbox Output Shaft

The gearbox output shaft and the leg connector are under extreme bending stress, having to transmit up to 35Nm of torque. Thus, it was highly important to ensure that the shaft, key-slot, shaft sleeve and leg connector can withstand the subsequent stress. The material of the shaft provided by *Matex* is a heat-treated steal, CM435, with a yield strength of 902MPa [81]. The sleeve (seen in Figure 5.8 a) is made from M300, a tool steel that can be heat-treated locally with a yield strength of 960MPa. The FEA of this assembly can be seen in Figure 5.8 and showed that the parts were strong enough. However, hand calculations followed to check the shafts FoS.

The compressive and shear stress between the shaft and the key were investigated as the shaft had a lower yield strength that the sleeve. By using the Jacobian inverse, the impact force is used to determine the torques seen by the shaft as if it was fixed in worst case scenario. A force of 1100N results in a torque of 129Nm at the shaft. The equations for determining key stresses are:

$$\sigma_c = \frac{T_{shaft}K_s}{hLr} = 660MPa \tag{5.2}$$

$$\tau = \frac{T_{shaft}K_s}{bLr} = 335MPa \tag{5.3}$$

Where h is the key height at 4mm, b is the key width at 4mm, L is the key length at 20mm

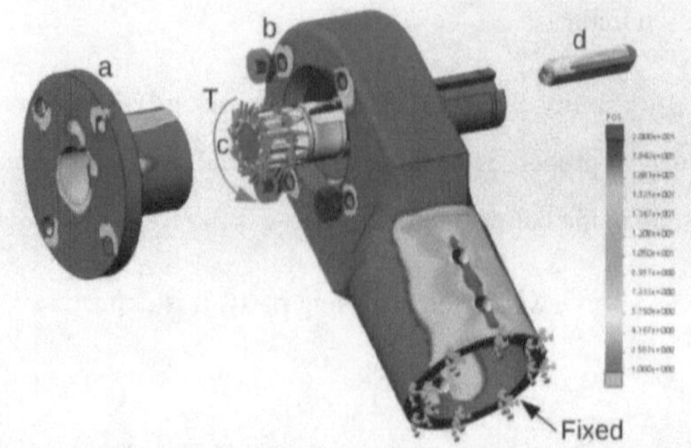

Figure 5.8: FEA analyses of the upper leg interface with the gearboxes with the heat map of the safety factor shown. The end of the leg connector is fixed representing a situation where maximum torque is output.

and r is shaft radius at 6mm. K_s (2) is the service factor determined from other factors such as shock loads, key tolerances and fatigue life factors. The calculated stresses were under the yield strength of the shaft and resulted in the FoS of 1.35. This is less than the desired FoS however, there were no other shafts that could be purchased from Matex. Furthermore, it was expected that this torque would be a dynamic load, with only a portion of the GRF resulting in static loading and the rest resulting in acceleration.

5.7.2 Other Parts

Aluminium is selected as the main material for the leg connector (as well as the other parts). A heat-treated alloy, 6083-T651, is selected as it is one of the stronger aluminium alloys, has good machining properties and has a price that is roughly the same as standard pure aluminium.

The other important parts that were considered less critical than the gearbox interface were reviewed to ensure they were strong enough. It was quickly seen that the common aluminium alloy 6083-T651 was also suitable for these parts. The leg support plate (figure 5.9a) had material removed to reduce weight but ensured that the key areas were thicker for strength. The foot was also analysed (see Figure 5.9b) and was seen to be more than strong enough to handle the expected ground reaction forces. The lower leg assembly was also tested in compression as per the expected ground reaction forces generated from the trajectory optimisation in Chapter 4. The minimum safety factor was found to be 2.

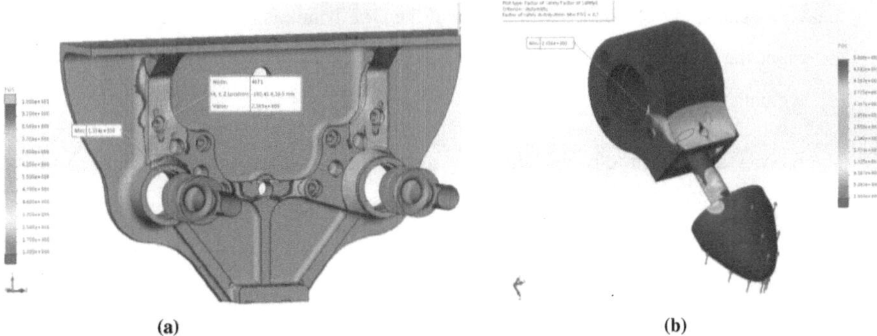

(a)   (b)

Figure 5.9: a) FEA of the leg support plate showing the heat map of the safety factor. **b)** The foot under the expected maximum ground reaction forces.

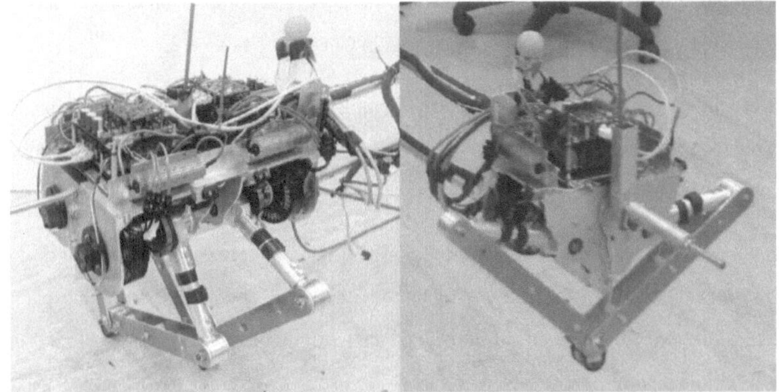

Figure 5.10: The final assembly of the bipedal and monopod, ready for operation as depicted.

5.8 Detailed Design Summary

The author compiled a drawing pack with all the required details to manufacture robot parts. The drawing pack for all the detailed components made in the UCT workshop can be found here along with several assembly videos that can be found here.

Once parts were manufactured, a single hip was initially constructed and any defected parts were sent back to the mechanical workshop. The fully assembled monopod and biped with all the required wiring can be seen in Figure 5.10.

Over all, the robot platform design was seen to be able to handle all the expected forces. The mass for a single leg was 6.8kg with the use of the U12 T-motors. This increases to 7.7Kg with the use of the U13 T-motors. Similarly, the biped weighs 14.1kg with the U12's and increases to 15kg with the U13's.

The robot was designed to be as lightweight as possible, however, ended up weighing 14.1kg. This was difficult to compare against other platforms as it is the mass specific-force that is most important. Part sizes were reduced as far as seen feasible with aluminium used as the main material.

The legs of the platform weighed a mere 13.32% of the body weight. This is similar to that of the MIT Cheetah's leg which was roughly 10% of the body mass and was seen to be negligible when modelling a legged robot's dynamics [36]. A final comparison between this robot's properties and other existing robots can be seen in Chapter 8, Table 8.2.

To ensure the design was significantly cheap, complicated parts were avoided such as the originally planned three splined shaft seen in Figure 5.4. Readily available aluminium sections and plates were used as the leg limbs so they can be easily replaced should they get damaged.

The robot platform was easy to assemble from the ground up. However, the author discovered that, in order to swap out the gearbox or remove the legs, separation of both hip plates was required along with removal of all embedded system components. This was found to be a tedious, time-consuming task.

The embedded system components were accommodated for, however, the author failed to adequately accommodate for all the wiring and how to secure it. This resulted in the platforms interior looking aesthetically displeasing.

The legs of the biped were modular in design with numerous bolt holes on the upper plate of the hip to allow different connections. Furthermore, the upper plate is a cheap aluminium sheet that can be cut in different manners, should it be required, providing provision for numerous additions in years to come. The foot design also allowed different connections to be machined and added if needed.

Recommendations for the second iteration of this platform are described in Chapter 9.

Chapter 6

Controller Design

The purpose of this chapter is to select and design a controller for the biped robot followed by simulations and verification of the controller in a physics engine.

With the advances in control structures, there are numerous new methods to model bipeds with more detail, taking into account the system dynamics. However, designing such a controller was outside the scope of this project and will be taken up by fellow PhD students in the future. For the purpose of platform verification, the Raibert controller was used as selected in the literature review (in Section 2.6.2).

6.1   Raibert Controller

Raibert defines how running and hopping for a monopod could be decomposed into three simple parts, avoiding platform modelling and making use of the natural dynamics of the system. These are:

- Height
- Forward velocity
- Body Pitch

This decomposition provided a framework of which a control system for bipeds and quadrupeds can be designed. Theses controllers rely on loose coupling between the different systems to provide sustainable locomotion.

Raibert identified that the legged locomotion could be represented by the Spring-loaded Inverted Pendulum (SLIP) which generated gate patterns very similar to those seen in nature [3]. This provided a simplified model to control. This project applies the above controllers to the SLIP model and then breaks the required actuation into the joint space (shown in Section

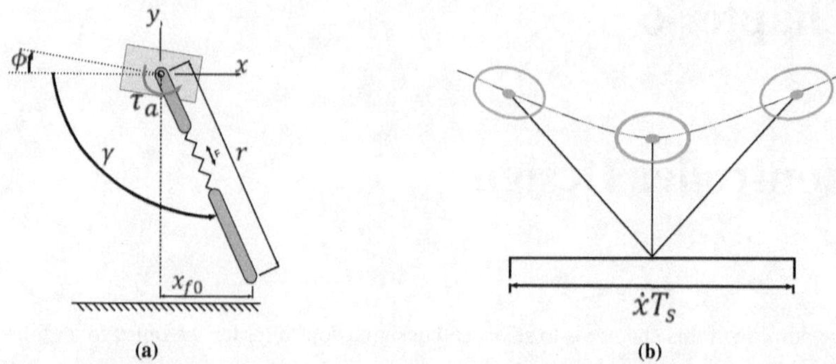

Figure 6.1: (a) The general SLIP model with the associated generalized coordinates showing the location of the neutral point (x_{f0}) relative to the body's centre of mass. ϕ is the hip angle, γ is the absolute leg angle, r is the leg length, τ_a is the leg angle torque and F is the prismatic leg force. **(b)** SLIP trajectory undergoing a steady state stance phase showing the length of the stride as the multiplication of forward velocity and stride duration.

6.1.2) of the linkage mechanism. The three 1 degree of freedom (1DOF) decoupled controllers mentioned above are applied during different states of a finite state-machine.

Height controller: The hopping height controller was used to regulate the amplitude of hopping and the main role was to input energy into the system to over come the losses due to inelastic collisions and the impedance of the leg. It was extremely difficult to determine the relationship between the thrust applied and the hopping height that would be achieved. Rather, Raibert suggests selecting a suitable thrust force, given iterative experiments to determine a sustainable hop. For this project this was done in the robot simulator V-Rep which uses the Bullet physics engine.

Forward velocity controller: The forward velocity controller regulates the velocity by determining the foot touchdown in front of the COM (see Figure 6.1a (a)). The neutral point (x_{f0}) for the stance is in the middle of the COM stride (see Figure 6.1b (b)) and this can be calculated knowing the forward velocity and stride duration.

Touchdown of the foot at x_{f0} results in a net forward acceleration of zero due to the symmetry of motion of the SLIP model. However, if the foot gets placed behind x_{f0}, it causes asymmetric motion that results in a positive net acceleration of the model and visa versa. Knowing this, a simple Proportional (P) controller seen in equation 6.1 was used to calculate the desired foot position (x_f), given the desire to either speed up or slow down [3]. Equation 6.2 then calculates the desired leg angle (γ), given the current leg length (r). The desired leg angle (γ) was controlled by a Proportional and Derivative (PD) controller in equation 6.3 that was tuned with the physical platform (see Chapter 8).

$$x_f = \frac{\dot{x}T_{stride}}{2} + K_{\dot{x}}(\dot{x} - \dot{x}_d) \tag{6.1}$$

$$\gamma = \phi - asin\left(\frac{\dot{x}T_{stride}}{2r} + \frac{K_{\dot{x}}(\dot{x} - \dot{x}_d)}{r}\right) \tag{6.2}$$

$$\tau_a = -K_{p\gamma}(\gamma - \gamma_d) - K_{d\gamma}\dot{\gamma} \tag{6.3}$$

Pitch controller: The body pitch was controlled with a simple PD controller that simply actuates the hip when the foot was in contact with the ground. This enables the body's orientation (θ) to be controlled by using the foot frictional force [3] to apply a torque to the body.

$$\tau_a = -K_{p\theta}(\theta - \theta_d) - K_{d\theta}\dot{\theta} \tag{6.4}$$

6.1.1   Expanding to biped controller

The above mentioned controllers could easily be extended to bipedal robots, however, it has to be ensured that only a single leg is in contact with the ground at any given time. Thus, the robot still behaves like the monopod [3]. The main difference was to achieve symmetry between the bipeds legs to reduce the net torque applied on the body. This was achieved by knowing that the forward leg angle is roughly equal but opposite to that of the hind leg, regardless of the forward speed [65]. To achieve symmetric leg swinging, the leg in the aerial phase was tasked with tracking the leg in contact with the ground. When both legs were in the air, the velocity controller (described above) took over to position the foot accordingly before touchdown. A PD controller in Equation 6.5 was used when leg 2 was making ground contact and thus, leg 1 tracks the opposite angle of leg 2.

$$\tau_{leg1} = -K_p(\phi_1 - (-\phi_2)) - K_d(\dot{\phi}_1 - (-\dot{\phi}_2)) \tag{6.5}$$

In addition, the leg lengths of the biped need to be controlled such that during one leg's stance, the other leg is retracted to avoid touching and tripping on the ground. This was achieved by modifying the neutral length (L_n) of the virtual compliant model described in the subsection below.

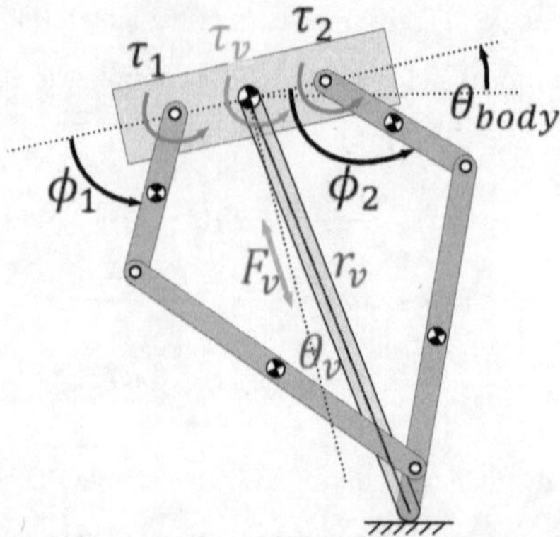

Figure 6.2: The five bar linkage model as the anchor to the Spring-loaded Inverted Pendulum model with the generalised coordinates and actuated joints of both shown. The jacobian transpose (J^T) is used to determine the required anchor torques (τ_1 and τ_2) from the SLIP template's torque/force (τ_v/F_v).

6.1.2   Virtual compliance

Given the use of the Spring-Loaded Inverted Pendulumn model, the radial length of the leg was modelled as a prismatic spring. This was accomplished by first choosing some neutral length (L_n), a spring constant K_l (N/m) and damping coefficient C_l (Ns/m) and then using the robotic leg's radial position and velocity to generate the desired force. In addition, when thrust force was required (needed for the Raibert controller), it is added as seen in Equation 6.6 by F_{thrust}. Should virtual compliance be needed for leg angle, the desired torque could be calculated in a similar manner to the prismatic spring (in Equation 6.7).

$$F_v = -K_l(r_v - L_n) - C_l\dot{r}_v + F_{thrust} \tag{6.6}$$

$$T_v = -K_{\theta_v}(\theta_v - \theta_n) - C_{\theta_v}\dot{\theta}_v \tag{6.7}$$

To achieve virtual compliance, such as if an actual spring and damper were connected to the leg, the control loop above needed to actuate the motors at near 1kHz time scales. Shown in Chapter 7, a time-scale of 500Hz was sufficient. The anchor (the underlying object that fixes the SLIP model to the physical world) for the SLIP model was the five bar linkage system shown in Figure 6.2. This anchor is used to transform the desired virtual forces and torques into the joint space. The Jacobian transpose of the scissor linkage transforms the desired virtual leg force (in Equation 6.6) into the required motor torques.

$$\begin{bmatrix} T_1 \\ T_2 \end{bmatrix} = J^T \begin{bmatrix} F_v \\ T_v \end{bmatrix} \tag{6.8}$$

The definition of a Jacobian is explained in Section 2.3.4.

6.2 Sequential state-machine

The sequential state machine was used to determine which controllers should be used through-out the motion and what events trigger the switching. The state machine increases in complexity for the biped compared to the monopod.

6.2.1 Monopod

The monopod state machine can be seen in Figure 6.3, which includes references to the controllers described in Section 6.1 above. The paragraph below is a description that accompanies the state machine diagram (see Figure 6.3). It should be noted throughout all the states, the leg length and angle were controlled via virtual compliance described in equation 6.6 and 6.7 by modifying the neutral spring length (L_n) and angle (θ_n). Furthermore, the spring constants were adjusted for the different states when needed such as increasing the stiffness and damping of the virtual spring upon landing.

Starting in the **flight state** for a monopod, the forward velocity controller calculates the desired foot touch down position. When contact has been made (positive ground reaction force), the state switches to the **contact state**. This phase provides a short period of time (compTime) to allow the normal force to increase before torquing the hip. This was crucial as it avoids the possibility of the end affecter slipping [3]. Note that in this state the spring constant and damping was increased to absorb the change in momentum of the robot. During the **compression state**, the attitude controller runs and torques the hip to remain horizontal. In addition, the monopod goes through the natural dynamics of the virtual linear spring (as in the SLIP model). The **thrust state** follows when the body reaches the trough of the stance phase. Here, a radial foot force was commanded additively to the virtual spring force in order to add energy into the

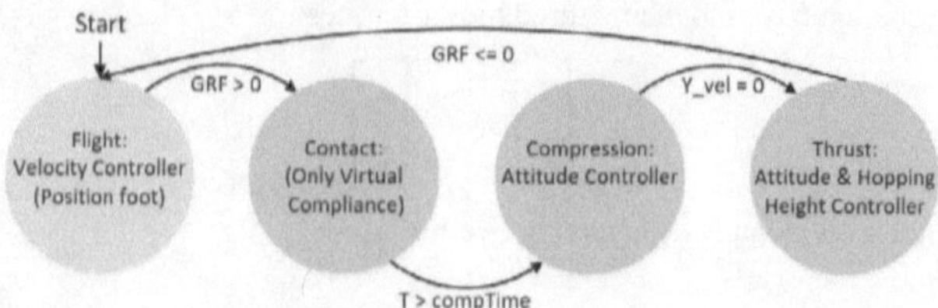

Figure 6.3: The state machine diagram for the monopod Raibert Controller.

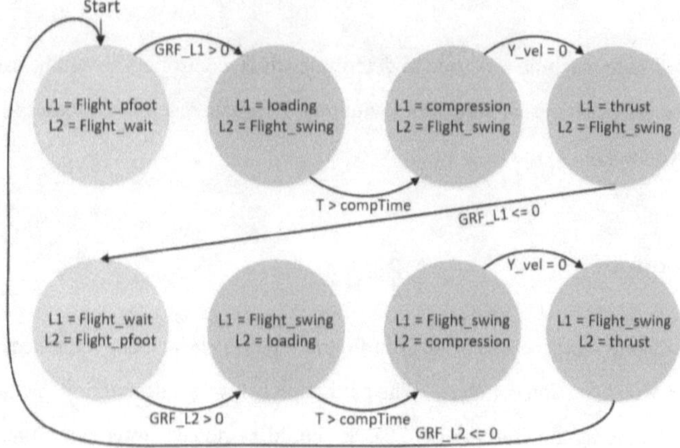

Figure 6.4: The state machine diagram for the Biped Raibert Controller.

system. When the monopod loses contact with the ground, the platform proceeds to start in the flight phase. The pseudo-code structure used to run the monopod can be seen in Appendix C.

6.2.2   Bipedal States

The bipedal state machine requires coordination between both legs. To achieve bipedal locomotion, the state machine seen in Figure 6.4 was used. In addition, several different controllers were added to allow for coordination of the legs. In each state, a leg was assigned one of six sub-states. Those are:

- **Flight_position_foot** - the forward velocity controller (Eq. 6.1, 6.2 and 6.3)
- **loading** - allow loading of the foot (before torquing hip)
- **compression** - attitude controller (Eq. 6.4)

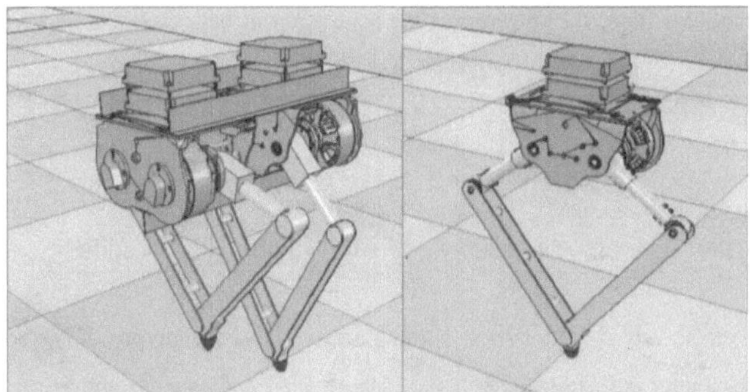

Figure 6.5: The biped and monopod models used in the physics engine V-Rep to simulate the Raibert Controller. A video of the simulations can be found here.

- **thrust** - attitude controller and added thrust to virtual compliance controller
- **Flight_wait** - leg remains at lift off angle but radially retracts
- **Flight_swing** - leg tracks the opposite angle of the leg contacting the ground (Eq. 6.5).

6.2.3   Position Control

The Raibert controller's commanded value was the forward velocity $\dot{x}_d$ seen in equation 6.1. This could simply be expanded to PD position controller where the new command input became a desired position value, x_d.

$$\dot{x}_d = K_{px}(x - x_d) + K_{dx}(\dot{x}) \tag{6.9}$$

6.3   Simulation Verification

To verify the suitability of the controllers for use on the monopod and bipedal platform, both controllers mentioned above were coded in Lua and implemented in a physics engine, *V-Rep*. The model's masses and inertias were taken from the SolidWorks design and can be seen in Table 6.1. This simulation allowed for quick iterative changes of controller gains and this gave an estimation of the gains that were needed for the physical robot (see Chapter 8).

During the initial hopping tests, the controller was very unstable for position control. For the

Table 6.1: Table describing the properties of the SolidWorks model's mass, inertia, length and width.

	Mass (kg)	Inertia ($kg.m^2$)	Length (mm)	Width/Diam (mm)
Femur	362.54	$1.9e^{-9}$	174.5	25.4
Thigh Narrow	375.72	$3.89e^{-9}$	295.5	34.6
Thigh Wide	205.7	$1.79e^{-9}$	300	43.8
Single Hip	5526.60	$4.13e^{-9}$	220	153
Both Hips	11599.4	$8.92e^{-9}$	220	400

robotic platform to run at higher speeds, the gain for the leg angle position (seen in equation 6.7) had to be increased to enable a higher leg swing frequency. However, at lower speeds it caused the system to be unstable, resulting in the legs overshooting small displacements required to stay in the same spot. Thus, to counteract this, gain scheduling was used. Should the commanded velocity be below $0.5m/s$, then the leg angle gain would be reduced.

With the above implemented and by tweaking the gains, stable limit cycles were achieved for both the monopod and biped. It can be seen in Figure 6.6 that the bipedal model was able to move to and maintain a commanded position. Furthermore, the figure shows the oscillations of the Z height and the ground reactions forces experienced by each leg. The stable Z limit cycle from the simulations can be seen in Figure 6.7. A video of the robots motions can be seen here.

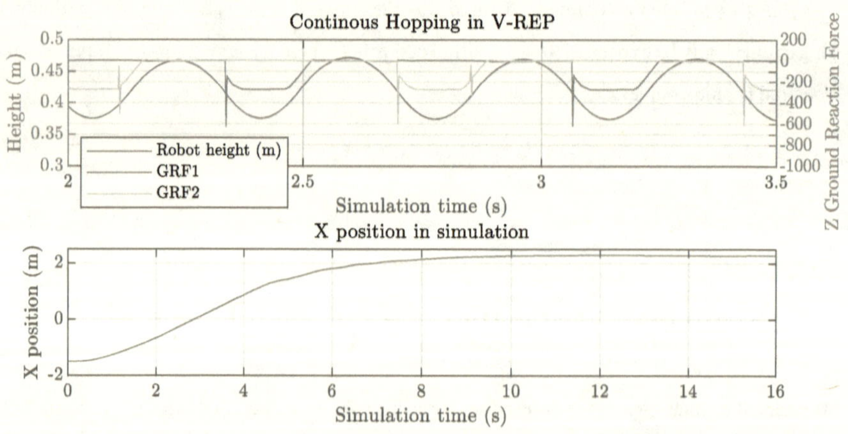

Figure 6.6: The top graph depicts the ground reaction forces from each foot in the biped (GRF1 and GRF2) and the body height on a short section of the simulation. The lower graph depicts the change in X position of the body over the entire duration of the simulation.

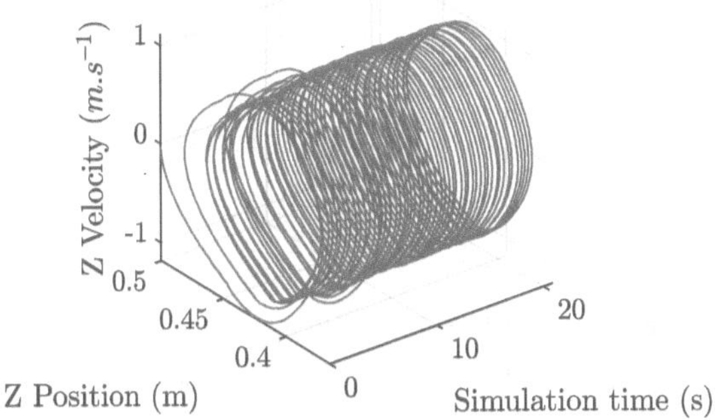

Figure 6.7: The Z limit cycle for the biped during a position control simulation.

6.3.1 Sensor inputs

After testing the controller, it was important to identify what sensor data was required for the implementation of the physical platform. This information was used during the embedded system design. The data required was:

- X position
- X velocity
- Z velocity
- Ground contact sensing
- Leg angles and velocities
- Body attitude

The sensors used are defined and configured in Chapter 7.

6.4 Summary

This chapter developed the Raibert control architecture that would be suitable to operate the bipedal legged robot. Furthermore, the controllers were testing in a physics engine to verify their suitability, but also assist in the initial selection of suitable gains for the physical platform.

Chapter 7

Hardware and Software Development

This chapter details the design and development of all the hardware and software aspects of the project needed to run and verify the robot. This covers the selection process and set up of all the electronic components used, the support method to ensure the robot is constrained in the sagittal plane, the sensor's selected and lastly the hardware and software implemented for the control scheme seen in Chapter 6.

7.1   Sagittal Plane Support

Limiting the robotic platform to motion only in the sagittal plane was one of the design limitations. This support system was developed first as it was essential for any experiments that were to be performed on the robot.

As seen in the previous chapter, limiting the robot to only the sagittal plane allows for motion in the X and Y direction with rotation around Z (see Figure 7.1). A single force, F_z along with two moments, T_y and T_x should be resisted to ensure the robot is constrained to the XY plane.

Several support systems concepts were considered. The main concepts identified were:

- The Parallelogram Boom [82][83];
- The rigid boom with linear bearings, and;
- The sled (or trolley).

The concept for each support system can be seen in Figure 7.2.

Rigid Boom

This boom design utilises a fixed boom around a rotary joint and linear bearings to allow the robot to jump up and down. The rigid Boom design has to support its own weight, reducing

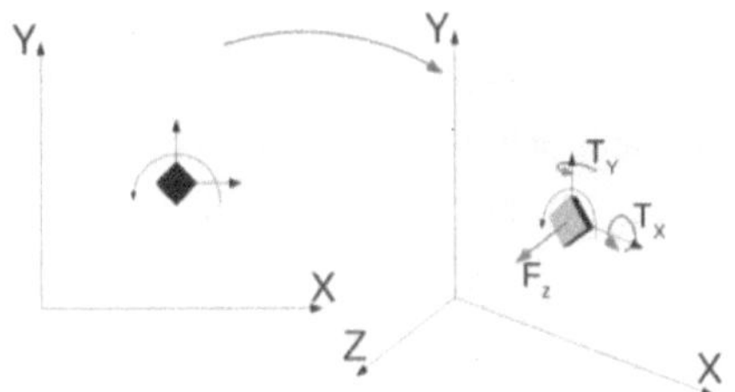

Figure 7.1: The left figure represents the plane the robotic platform should exist in. The red arrows in the right figure indicate the forces and torques that the support would have to resist in order to keep the biped robot in the desired plane.

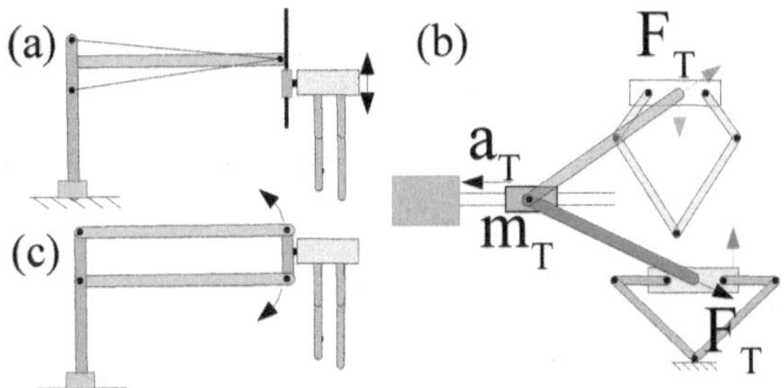

Figure 7.2: **(a)** The rigid boom concept. **(b)** The sled concept and the detrimental forces due to the sliding mass. **(c)** The parallelogram boom concept.

the load the robot would have to carry. However, this creates an asymmetric inertia affect on the robot. In the vertical plane, the robot feels a certain small added mass (the linear locating bearings), but in the horizontal plane the robot would feel the inertia affects of the entire boom mass that has to be accelerated. In addition, the complexity of ensure the design can hold its own weight is undesirable.

Parallelogram Boom

This boom uses a four bar linkage that ensures that the robot always remains perpendicular to the ground, regardless of its height (such as during jumping). The only concern is that as the boom arm increases in height, the projected length of the boom on the ground gets shorter, potentially placing out of plane forces on the robot's leg as it leaps upwards. From rough calculations with a realistic boom length of 2.2m, this offset was determined to be roughly 10mm at maximum which was considered negligible. Furthermore, this design acts simply as an added mass for the robot.

Sled

The sled (or trolley) support system requires a lot of infrastructure to implement. It involves designing a trolley or sled that is pulled along by the robot in a linear motion and requires a rail system for the trolley to sit on. The trolley holds the robot in a plane and allows vertical motion.

Boom Selection

From the existing designs above, the sled was immediately eliminated as a possibility. The system was the most complex design presented and would require a significant amount of time and money to design without severe issues. One such concern was that if there was a significant amount of friction in the linear bearings of the sled, which may arise due to slight misalignment over the hopping distance, the robot would have to work against that friction. In addition, a simulation was run in a physics engine, V-Rep, simulating this support. The results quickly indicated that the up and down motion of the robot body would cause the sled to push and pull the torso, creating a highly unstable environment for the biped. This is visualised in Figure 7.2 (b).

The Rigid Boom was desirable for significantly reducing the weight the robot must carry. However, the asymmetric affect of added mass (robot feels added mass in the tangential plane) of the boom along with the complexity of the design to support its own mass made it undesirable. Furthermore, numerous parts would be needed to locate the boom arm, increasing the mass of the boom.

The Parallelogram Boom was ultimately chosen as it required minimal, simplistic parts to be manufactured along with the symmetric affects of added mass on the robotic platform. The final boom design in SolidWorks can be seen in Figure 7.3. To measure the attitude of the robot, an encoder was incorporated in the design of the boom. The mass of the boom end affecter seen by the robot was 1.42 kg. Simulations were also run in V-Rep to identify the expected forces and torques required to hold the robot in place. This was to ensure that the boom was strong enough to withstand the expected external disturbances. The simulated torques from the biped robot hopping (seen in Section 6) and the foot height above the ground can be seen in Figure 7.5.

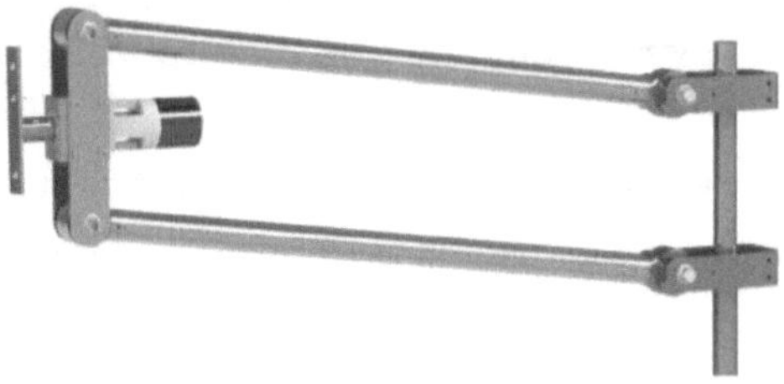

Figure 7.3: The final SolidWorks design of the boom along an image of the final assembly. The weight of this boom felt by the robot was 1.42kg.

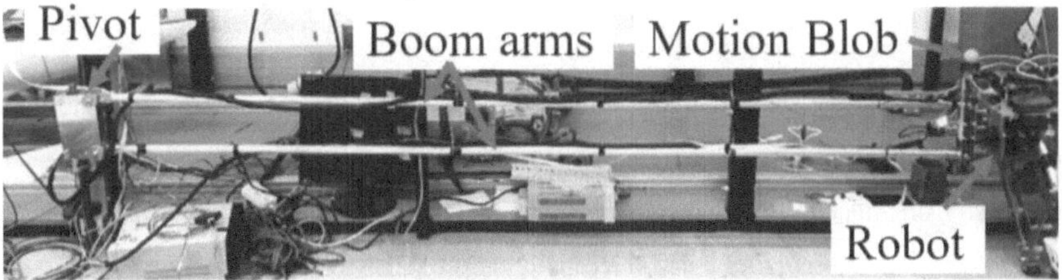

Figure 7.4: The final assembled boom with the motion tracking blob (explained further in this chapter) and the robot attached.

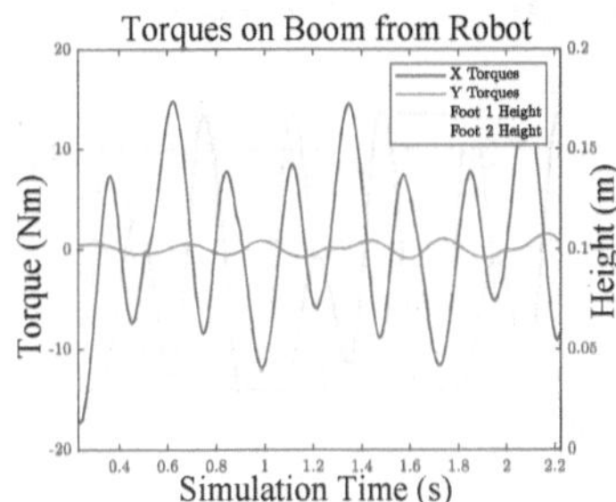

Figure 7.5: The torques expected to be applied to the boom arm during typical biped hopping.

7.2 Embedded System

To verify the mechanical platform, the embedded system was set up to control all the actuators and sense the desired properties needed by the controller described in Chapter 6. This section describes the selection and implementation process.

7.2.1   Controller Environment and Communication protocol

In the past, most robotic platforms used micro-controllers to perform the brunt work for controlling motors, performing the calculations for the controller and bringing in all the sensor data. However, with increasing components, the shear number of communication wires can decrease the reliability (by increasing the system complexity). Furthermore, common micro-controllers also lack the processing power to execute complex control strategies at near 1KHz speeds.

A solution to this was the EtherCAT communication protocol. This uses standard Ethernet cables to daisy chain slave devices together and allows all slave devices to communicate with the master simultaneously. This protocol is plug and play, allowing devices to easily be added or removed. These properties dwarf the capabilities of standard communication methods such as RS232, I2C and SPI. For example EtherCAT transfers data at rates up to 100Mbit/s [84].

For this very reason the EtherCAT protocol was selected with longevity and improvements in mind. This relates to the overarching project that will see the embedded system reused for many years into the future for different iterations of the mechanical platform. This premise should remain with the reader throughout this section as numerous expensive components are purchased with the future and re-usability in mind.

The Simulink Real-time environment was selected as it is able to initialise and use the EtherCAT bus. The controller for the robot was designed in Simulink and was compiled and pushed onto the target PC as a C++ program that runs at real-time. Furthermore, Simulink Real-time runs directly on the CPU, which in this project is an Intel i5 CPU. This provides more than enough processing power to run at speeds up to 1KHz and perform complicated calculations for the controller.

7.2.2   Motor Drivers

The EtherCAT protocol requires that all slave devices are EtherCAT compatible which guides the motor driver choice. With the assistance of both the author's Supervisor and a PhD student, the motor drivers were selected to be Ingenia Jupiter Drives. These motor drivers are capable of driving up to 80A peak and 40A continuous. These ratings are suitable for the U12s and U13s used on the robot. The motor drivers are rated for network commands up to 1000Hz, however, during the configuration and testing of the control system, this speed was not achievable. Thus, the real-time operating system frame time had to be lowered to 500Hz (which was identified during testing in Section 8.1).

To ensure the safety of the system, the Ingenia Jupiter Drives have an input boolean pin that can kill the drivers at any time. Limit switches for the hard stops of the robot are connected to this pin. If the upper link of the robot leg hits a hard stop and activated a limit switch, the motor drivers would be killed. The hard stops can be seen in Chapter 5, Figure 5.6.

7.2.3 Required sensors and Inputs

The controller described in Section 6.3.1 indicates what sensor inputs were needed along with other desirable inputs to manually control the robot:

- Forward velocity;
- Vertical velocity;
- Motor position and velocity;
- Body attitude angle;
- Displacement (optional for position control);
- Ground contact sensor (for hybrid state machine), and;
- Buttons to initialise, start and kill the control system.

To enable external control of Simulink Real-time system while it was running, such as activating initialisation procedures, switching states and having a kill button, the EK1818 Beckhof terminal was purchased. This terminal incorporates external digital boolean signals into the EtherCAT protocol packet. Furthermore, this unit allows other terminal blocks to pass data into the EtherCAT bus (as needed by the terminals below).

Standard encoders (500 counts per revolution) were chosen for measuring the motor angles and were designed for, as seen in Chapter 6. The motor drivers selected incorporate the motor encoders into the EtherCAT PDO (process data objects) that were sent to the master. The body attitude was determined via an encoder located on the boom arm, as seen above in Section

7.1. However, to view the body encoder values, a Beckhoff terminal block (EL5152) was purchased and connected to the EK1818 block. This device can take in several encoder outputs and adds a 32bit number into the EtherCAT protocol packet representing the encoder ticks. Two six-axis force sensors (ATI Axia 80F/T Sensor) were purchased to enable ground contact detection for the sequential state machine and to measure the ground reaction forces of the robot. Furthermore, these sensors were required for use by a PhD student in future work. Additional Ethernet ports were needed for the force sensors, therefore, another terminal was purchased for this purpose (the EK1122).

Lastly to read the forward velocity of the robotic platform there were two different options. The first option was to use an encoder at the centre of the boom to measure the angular position of the robot and integrate that to find the forward velocity knowing the boom length. However, given the poor mapping from the encoder angle to arc position (due to the large radius of the boom), it was decided to use an existing camera system. This system was able to track infrared blobs (ping pong balls with infrared LEDs) within the experimental space (similar to the well known Vicon Motion System). This was originally designed by a PhD student[1] for a quad-copter. The system tracks several blobs and is able to calculate the X, Y and Z position and velocities using an Extended Kalman Filter [73]. For the purposes of this project the number of tracked blobs was reduced to one and the author set up a dedicated Linux machine with a GUI to send the position and velocity data via RS485 to the Simulink Real-time System using the Beckhoff EL6022 terminal block. The location of the blob on the robot can be seen by the orange pingpong ball in Chapter 5, Figure 5.10 or in Figure 7.4. The GUI's for the motion capture system and host computer along with images of the target computer and the EtherCAT terminals can be seen in Appendix D.

7.3 Summary

The work completed within this chapter provided the frame work for enabling the testing of the robotic platform. Several different sagittal plan support methods for the robot were investigated, with the most suitable design, the parallelogram boom, designed and assembled.

---

[1] Acknowledgement of Arnold Petorius, a PhD candidate in the Mechatronics Research Lab, who developed the visual tracking system.

The motor drivers were configured for the attached motors. The Simulink Real-time operating system and host computer were set up to run the various controllers required. In addition, the required sensors for the controllers were selected and developed. The EtherCAT protocol was used to communicate between the real-time system and other devices. Any devices that were not compatible were connected via several EtherCAT terminal blocks. This resulted in a fully functioning sensor suite and control system that was read to operate the robot.

Chapter 8

Platform Experiments and Verification

This chapter details all the experiments, covering the results for the single modular leg and the biped. To provide a thorough review of experiments, this chapter will group each individual experiment together with the relevant results and discussion. At the end of this chapter a summarised discussion is presented covering all notable results. An interactive Simulink Control structure used for all experiments can be seen here.

8.1   Initial Robot Testing

For initial tests, the single leg module was mounted to a prismatic test rig (see Figure 8.1 (a)). Before full hopping tests on this rig began, the leg was initially commanded to act as a spring damper system using active compliance (see the function used in Section C.1). The output torque was limited to 10% of the rated value.

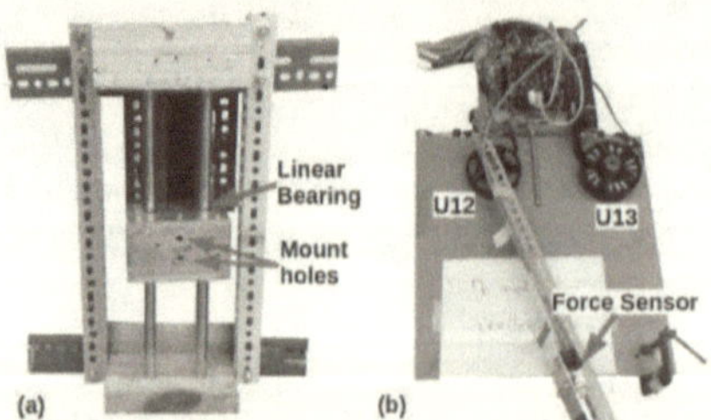

Figure 8.1: a) Prismatic support for initial testing of the robotic leg. **b)** Motor comparison test setup.

Several bugs occurred during initial testing. The motor drivers would stop responding spontaneously while all other systems continued to function. This was extremely dangerous as on more than one occasion, the motors would drive the output linkage into the hip of the robot. A process of elimination led to the cause. The motor drivers and EtherCAT protocol system were configured to send data at 1kHz (as per the rating of the motor drivers), however, adjusting this down to 500Hz removed the above fault completely. This issue resided within the Ingenia motor drives as all other sensor data continued to function while the drivers stopped responding. It was out of the scope of this project to fix this issue as it seemed to be an internal fault on the boards. Beyond this no other major hardware issues occurred. This issue was reported to the Ingenia support team and they have not yet solved the problem.

8.1.1 Motor Performance

To identify the start-up torque, the U13 T-motors had a torque ramp applied to each motor of the leg. The set point was to keep the femur links horizontal and apply the same torque ramp magnitude by performing tests with both positive and negative increasing torques. This was to remove the torque bias due to the weight of the lower links. It was clearly seen in Figure 8.2 that the start up torque for the U13 T-motor was significant with an average of $0.9Nm$ for motor 1 and $1Nm$ for motor 2. By working with the motors, it was evident that this was due to the cogging affect seen in brushless DC motors. Cogging is the torque generated by the magnetic attraction between the permanent magnets in the rotor and the iron cores of the stator [85].

Table 8.1: U12 and U13 T-motor properties are compared along with the difference in metrics in the right most column. The highlighted green cells indicate improved metrics of the U12 over the U13, most notably the Mass specific torque output.

	U12	U13	Diff
Torque Constant (Nm/A)	0.106	0.112	0.006
Peak Current (A)	100	160	60
Mass (g)	794	1280	486
Number of poles	36	24	12
Cogging Torque (Nm)	0.043	1	0.957
Rated Torque (Nm)	5.305	7.302	1.998
Max torque at Driver Limit (Nm)	8.48	8.96	0.48
Mass specific torque output (Nm/kg)	10.68	7	3.68
Foot Force @ Max Torque (N)	557.9	589.5	31.6
Mass specific Foot Force (N/kg)[a]	≈ 82	≈ 79	3

[a]The Mass specific foot force is calculated using the inverse jacobian transpose $((J^T)^{-1})$ of the leg configuration in the middle of the workspace divided by the mass of the robot with the different motors seen in Section 5.8.

This start-up torque due to cogging was roughly 15% of the rated torque of the U13 T-motors, 4.5Nm at the output. This was equivalent to 42.3N of resistance in the radial direction at the end

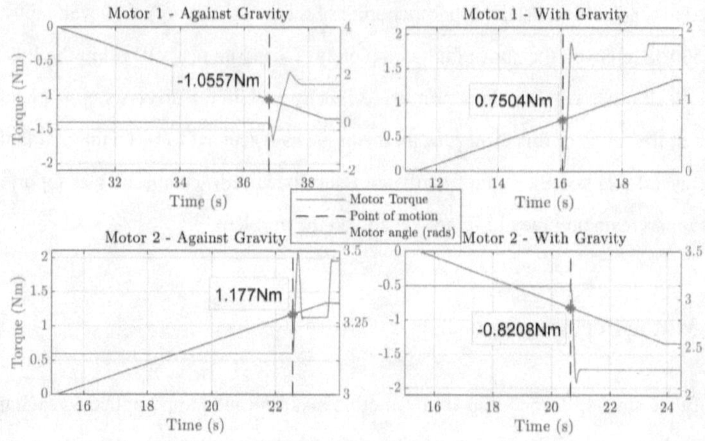

Figure 8.2: The startup torque for both motors while connected to the robot leg. The left figures are where the torque ramp opposes the torque generated by the weight of the femur and tibia links. The right column is where the torque ramp works with the weight.

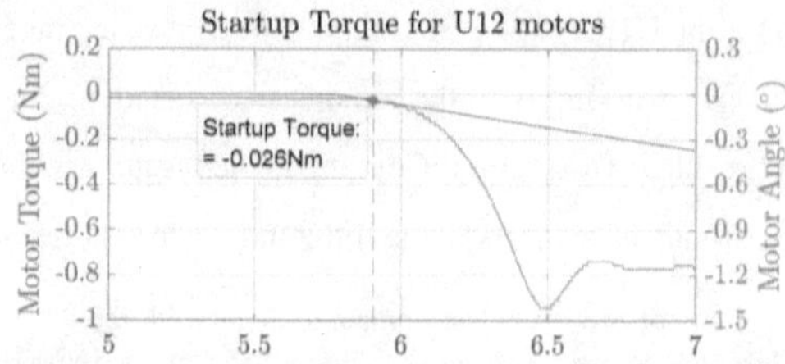

Figure 8.3: The start-up torque during one of the experiments where a torque ramp was applied to the U12 motor until motion occurs.

affecter. Such a high resistance was highly undesirable and not suitable for accurate position control, impedance control and proprioception.

The U13 motors have only 20 poles which directly increases the cogging strength [85]. There were numerous techniques, however, three of the four suggestions by M.S. Islam [85] were physical alterations to the motor and the last was an overlaying control loop. It was decided to rather investigate the U12 motors which were currently available which have 36 poles. A testing rig was built for both the U12 and U13 T-motors (see Figure 8.1 (b)). When investigating the start up torque of the U12 motors, it was seen to be on average a mere 1% (0.043Nm) of the rated torque. This torque ramp and initial motion of the U12 motors can be seen in Figure 8.3.

In addition, the output torque of both motors were compared while demanding a constant torque

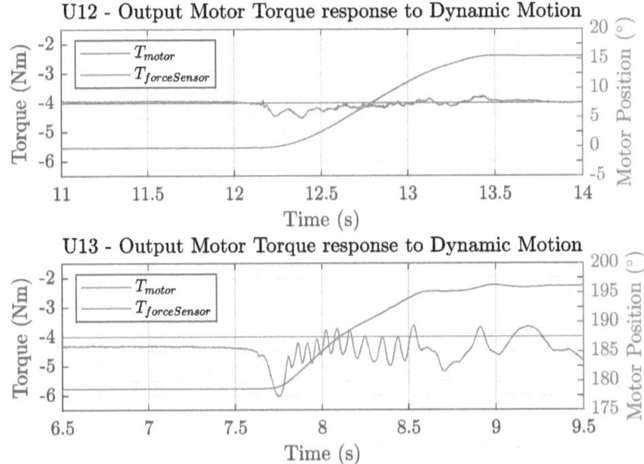

Figure 8.4: Both plots (upper for the U12 and lower for the U13) show the variation of the output motor torque while the motor was rotating compared to the commanded torque. It was clearly seen that the U12 performance was drastically better than the U13 which oscillates with amplitudes of 0.5-1 Nm.

and undergoing rotation. This was achieved by applying a force through the six-axis force sensor and rotating the motor via a lever arm of known length (see Figure 8.1). From this experiment the output torque of the motors can be calculated and shows the consistency of the torque throughout the motor rotation. The results of both motors can be seen in Figure 8.4. The U12 motor has minimal oscillations while the U13 motor has extremely large oscillations of up to 1Nm due to the cogging torque.

Other metrics between the two motors were compared which can be seen in Table 8.1. It was noted that the Jupiter Drives (seen in Section 7.2.2) can only provide a constant current of 40A and peak of 80A. With this limitation it was determined that the U12's were not only a viable replacement for U13s, but would allow the robot to perform better. They may come at a cost of slightly less torque output but the reduction of weight improves the mass-specific force output of the robot. From these results, the author adapted the mechanical design to use the U12 motors instead of th U13 motors.

8.1.2 Foot design

Once initial testing was complete, a hopping controller for a monopod was implemented and tested on the prismatic support (the controller design can be seen in Section 6.1). To identify a

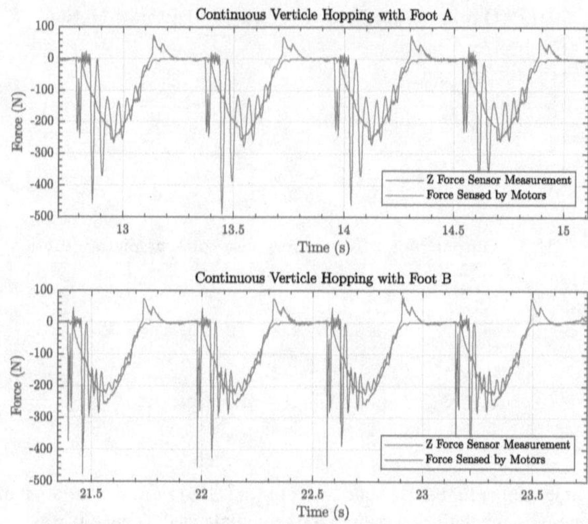

Figure 8.5: The foot force as measured by the force sensors plotted against the force seen by the motors. It was clear that the initial rubber damper foot caused severe oscillations that had a longer settling time. Furthermore the video seen here showed the foot causing the robot to oscillate out of plane.

suitable spring and damper constant for the controller template, a physics engine, V-Rep, was used to simulate the single leg (see simulations in 6.3). Several iterations were performed until suitable values were identified. In the experimental set up, a force sensor was located below the robot leg to measure the force generated during the compression and thrust phase of motion, as well as to control the switching of the sequential state machine.

The results of the first hopping test indicated that there was severe oscillations of the leg (see top plot in Figure 8.5). Visually, the robotic platform vibrated and through slow motion video (which can be seen in the video here), the foot design was identified as the cause.

This foot was a damper and from the results gathered, allowed significant deflection with a low stiffness. This was concluded to be the cause of the undesirable oscillating affects witnessed.

Given the foot blocks design to easily change foot connections, a new design was made with harder rubber to ease the oscillations on ground impact. The vibrations caused by the first design just after the initial impulse had an amplitude of $\approx 85N$ at a frequency of $24Hz$ and does not decay quickly. The new foot has peaks of $\approx 23N$ at a frequency of $38Hz$ and has a much higher damping. Thus, the amplitude decays much quicker. The new foot may suffer large oscillations at the start of the ground contact but performs better throughout the second half of

Figure 8.6: The different feet designs. On the left is the original foot (foot A) damper that caused significant oscillations on landing. On the right is the new foot (foot B) with a 3mm rubber pad, providing a smoother landing.

the ground contact phase. This was clearly seen in lower plot of Figure 8.5. The two different feet designs can be seen in Figure 8.6.

8.2 Force Transparency

Force transparency is highly desirable and to measure the accuracy achievable, the single leg module underwent several different virtual compliance tests with external disturbances. The actual motor torques (T_m), calculated using the motor current (determined internally on the Jupiter Drives), were multiplied by the inverse of the Jacobian transpose of the leg geometry to find the force at end affecter.

$$F_{feedback} = (J^T)^{-1} T_m \tag{8.1}$$

Any discrepancies between the force feedback from the motors and the force sensor underneath the foot are due to either geometric errors, impedance, motor cogging torque [85], impact forces or encoder initialisation offsets. To the best of the author's ability, the encoders are initialised as accurately as possible and the geometric discrepancies of the model are minimised by using the SolidWorks model. Thus, it was assumed that any offsets are purely the result of friction and impact affects.

It was clearly seen in Figure 8.7 that upon impact there was a large force spike followed by numerous small oscillations most likely due to the rubber foot of the leg. The maximum error occurs at these spikes and was offset from the force feedback by roughly 600N. However, if

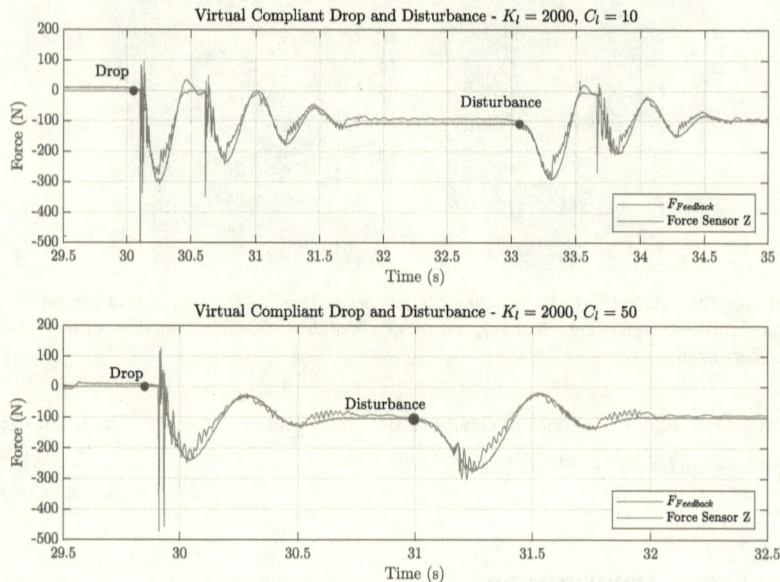

Figure 8.7: This data shows the single leg module undergoing virtual compliance testing where an external disturbance was applied and the platform was raised and dropped (each test seen by the blue marker). The purpose of this test was to identify the force transparency of the system by comparing the force calculated from the torque feedback (Eq. 8.1 used to calculate the $F_{Feedback}$) to the six-axis *ATI* force torque sensor reading.

the force sensor data was passed though a low pass filter, the maximum error spikes would be significantly reduced as they last for 2 to 4 ms (1 or 2 process frames). The average error percentage from the $F_{feedback}$ throughout the tests (seen in Figure 8.7) was significantly better with an accuracy of $\pm 16\%$. However, the performance was better represented by noting that the average error for the above tests was -6N with a standard deviation of 21N. This indicates that the leg mechanism and drive train was suitably designed for accurate proprioception, given the forces during a typical dynamic ground contact exceed 100N. A video of the leg under virtual compliance control can be seen here.

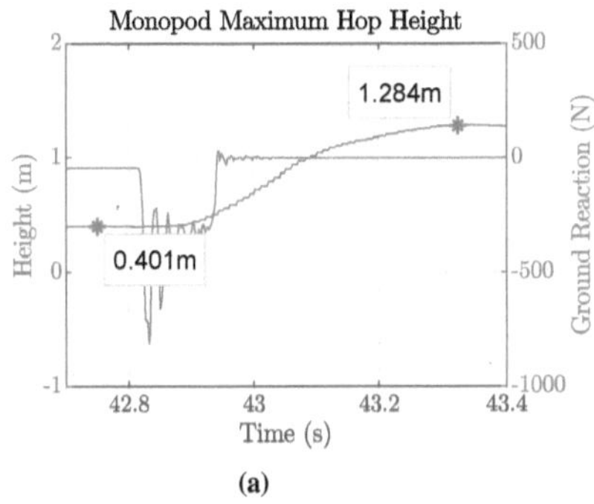

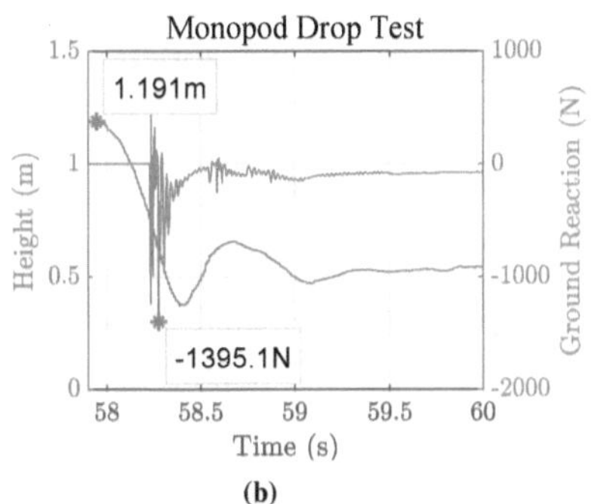

Figure 8.8: a) The maximum hopping height achieved by the biped showing the Z height and ground reaction force. The robot was caught via a bungee cord at apex of the leap **b)** Drop test from roughly the same height achieved during the highest leap. Shown are the Z height and ground reaction forces.

8.3 Individual Hop and Drop Tests

To ensure that the platform does not undergo any damage, it was decided that the hopping height would be investigated first. The maximum achievable height of the robot was then used in follow up drop tests to ensure that the platform can sustainably land from that height.

To perform the hop test, Baleka was mounted on the boom arm and located in place using a rope and bungee cord, limiting the leg to 1 DOF (Z height only). To perform the hop tests, the standard virtual leg template was used (as described in Chapter 6) and a foot force of $600N$ was commanded at the foot to saturate the motor drives. When the leg length of the template exceeds $0.45m$, the radial leg damping was toggled to $150N/ms^{-1}$ to slow the foot down and avoid self impact.

In Figure 8.8b a) the maximum achievable hopping height of a single leg was on average 0.81m, measured across three tests. The drop tests are performed using the same set up but the virtual leg template was given a spring constant of $1500N/m$ with the neutral point at almost full extension and a damping constant of $120N/ms^{-1}$ for landing. The leg was dropped at incrementing heights to ensure it could sustainably land. Ultimately, the platform was able to land from the maximum hopping height and the result can be seen in Figure 8.8b b).

8.4   Vertical Agility (VA)

To assess the agility Baleka, a metric was identified that has been used to evaluate existing robots and animals, providing a reference. Furthermore, vertical hopping is an important survival technique in nature for animals, such as hopping to catch prey or to escape predator. The metric defined by Duncin was Vertical Agility where the height a system can reach with a single jump (h) was multiplied by the frequency of repeated jumps (t_{stance} is the duration from zero velocity until take-off while t_{arial} is from lift-off until the apex of flight) [2].

$$VA = \frac{h}{t_{stance} + t_{arial}} \tag{8.2}$$

It should be noted that all legged animals are limited to a ballistic trajectory. Thus, there was a point where hopping height and frequency can go no higher. This is easily realised and seen in equation 8.3, where gravity limits the rate at which a system can return to the ground. Note that h is the vertical hopping height, τ is the hopping period and g is the gravitational constant.

$$\frac{2h}{\tau^2} \leq g \tag{8.3}$$

There are concerns that vertical agility was not a good indicator for the forward direction performance which was expanded on by Duperret [86]. He suggested a new metric called *Specific agility*. However, this concern was justified when investigating and comparing quadrupeds which rely mainly on the rear legs during rapid acceleration and suffer from other asymmetric issues. On the other hand, a biped's vertical agility can be considered a good metric for forward acceleration. This is because during a rapid acceleration manoeuvre, the entire robot's torso can lean forward and actuate with the legs as if it were a normal vertical jump. Thus, if a biped excels in vertical agility, it is a reasonable assumption that the robot would perform well in forward planar agility. The only limitation of the vertical agility test ultimately comes down to the friction at the foot which is irrelevant of the power of the platform.

Verticle Agility Results

As per previous experiments, the virtual leg template was used to control the leg. The hopping controller state machine for the monopod and biped described in Section 6.2 was used but the attitude and forward velocity controllers are removed. In addition, during landing, the

(a)

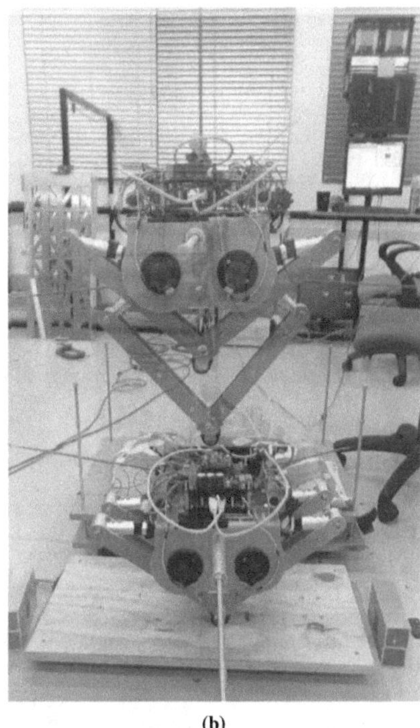

(b)

Figure 8.9: a) The monopod performing its maximum cyclic jumping restrained in 1DOF seen at the apex and trough of the motion. **b)** Baleka performing it's specific agility tests shown at the apex and trough of the motion.

controller lowers the body to its minimum position to ensure maximum explosive leaping. An image of the robot performing this motion can be seen in Figure 8.9a. The results of these experiments are tabulated in Table 8.2 which also compares Baleka's other properties.

The single leg was able to jump to an average height of 0.81 m (change in body height over five tests). Using equation 8.2, the vertical agility for the monopod was calculated to be 1.82 m/s. This value was best compared in a graph seen in Figure 8.10, where the vertical agility of the leg (both monopod and biped versions) are compared to animals and existing robots. It was clearly seen that the monopod outperforms all other robots except for the GOAT monopod which has a VA of 1.88 m/s. However, the designer [19] performs his experiments with no motor drivers or other peripherals on the platform, whereas Baleka has to jump with motor drivers and 1.42kg boom. Thus, it was expected that Baleka on its own would outperform the GOAT leg. The only animal seen to out perform this robot is the Galago which has a VA of 2.2

m/s.

The Biped configuration was tested in a similar manner to determine the vertical agility by jumping with a single leg while the other was held retracted. The platform was able to explosively jump to a maximum height of 0.54 m. During a follow up experiment, leaping with both legs simultaneously, the platform can reach a maximum height of 0.92 m which was higher than that of the single leg's 0.81 m.

As expected, when the biped jumped with a single leg, the vertical agility was significantly reduced from the 1.82 m/s of the monopod to 1.33 m/s. This was attributed to the increased mass but constant power output. Even with only a single leg used, the platform was able to completely outperform other assembled robots with multiple legs such as the ATRIAS biped robot with a VA of 0.22 m/s or the Minitaur quadruped with a VA of 1.12 m/s. The VA of the biped leaping with both legs was 1.86 m/s, slightly greater than that of the monopod configuration. The increased jump height and vertical agility of the biped compared to the monopod was attributed to the reduced added mass of the boom per leg, as all other parameters (both mass and power) are doubled. It was unclear whether the human test subject in Figure 8.10 jumped with one or two legs, nevertheless, the Baleka biped leaping with one leg has a higher vertical agility.

In addition, the maximum hopping height of the biped with a single leg is 0.54 m, greater than all other bipedal and quadruped robots. The closest was the Minituar jumping to 0.48 m. With both legs used at the same time Baleka was able to leap to 0.92 m, higher than that achieved by the GOAT leg. A full comparison of the performance metrics compared against other platforms can be seen in Table 8.2.

8.5   Continuous Hopping

Beyond performance testing, it was a requirement that the platform hops continuously to verify its performance for rapid acceleration manoeuvres and robustness. The controller described in Section 6.2 was used, but without the attitude and position controllers. To ensure the platform stays in plane during the hopping manoeuvres, the boom was secured such that the robot can only move vertically. The monopod configuration can be seen in Figure 8.11, which per-

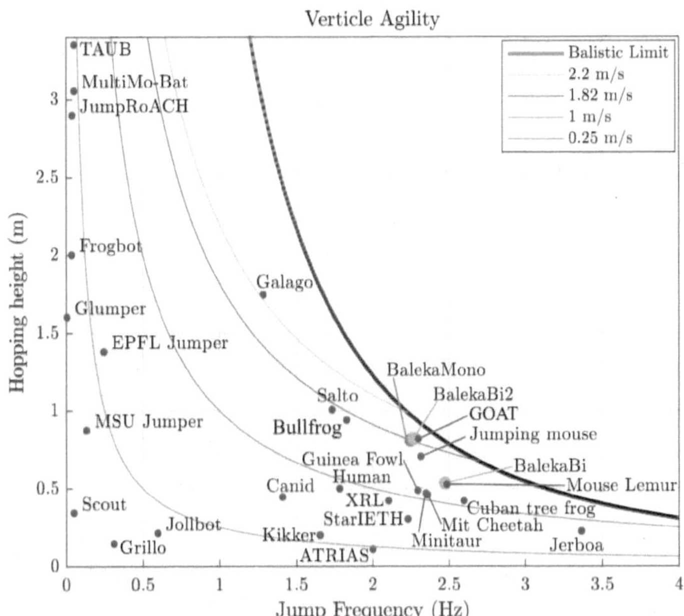

Figure 8.10: The vertical agility of known robots and animals compared to the platform developed in this work. The legend depicts various vertical agility curves including the ballistic limit. *BalekaMono* represents the monopod configuration. *BalekaBi* indicates the performance of the biped configuration but leaping off a single leg. *BalekaBi2* represents the biped configuration and leaping off with two legs. This graph was adapted from [2]

forms its maximum jump height. Figure 8.12 depicts information for Baleka during continuous hopping. A video of the biped hopping can be seen here.

The monopod ground reaction forces peak at around 850 N for the single leg. Comparing the 86 kg of force to the weight of the leg, 8.24 kg, the ground reaction force was over 10 times as large. This indicates that the performance of soft touch down and compliance was still far away from that achieved by humans which is only 4 times the bodyweight [33]. It was similarly seen in the video (see here), that the foot bounces of the ground at first touch down.

The ground reaction force for the biped similarly peaks at around 900 N and this was about 5.8 times the mass of the robot. This touch town peak is closer to that of humans [33]. The impact force spike was very similar between the biped and monopod configuration, even with the biped weighing almost twice as much. This indicates that the impact forces produced are

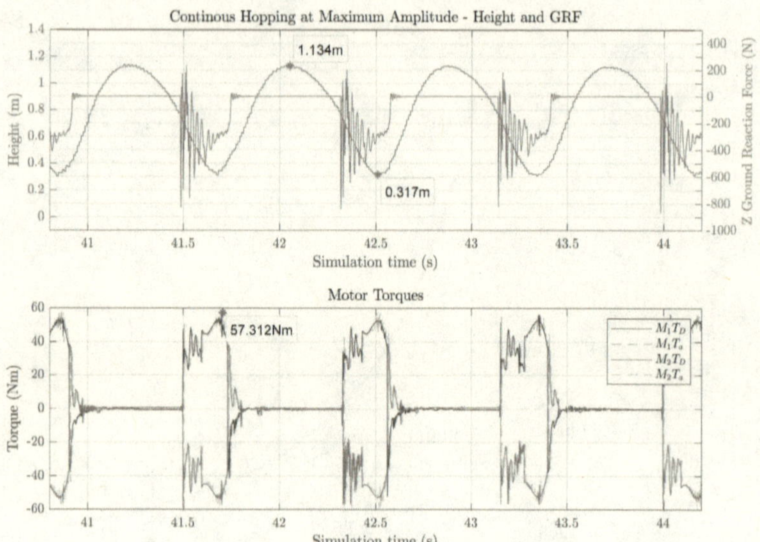

Figure 8.11: The single leg of the platform hopping at the maximum achievable height. In the top graph the body height and ground reaction forces are depicted while the motor torques are shown in the lower graph. The maximum change in body height is 0.919m from trough to apex.

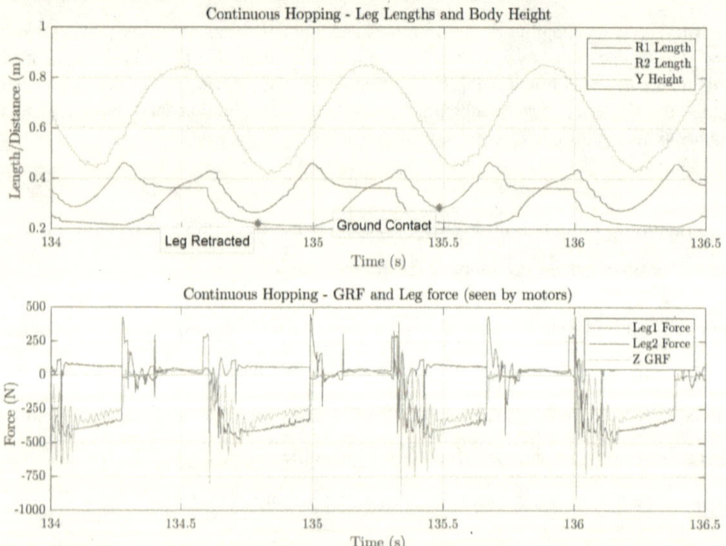

Figure 8.12: The biped continuously hopping on alternating legs with the top graph showing the body height and the leg lengths and the lower graph indicating the ground reaction force (force sensor) and the respective forces seen my the motors. For a video of the continuous hopping of both the biped and monopod please go here and here respectively. This hopping motion can be seen in the video here.

Robot name	Gear Ratio	no legs	leg length	Body Mass (kg)	Leg mass %	Motor Mass %	Max Jump Height (m)	hop freq (Hz)	vertical agility
BalekaMono	7	1	0.5	8.24	13.87%	23.28	0.88	2.25	1.82
BalekaBi	7	1	0.5	15.62	13.32%	22.37	0.54	2.47	1.33
BalekaBi2	7	2	0.5	15.62	13.32%	22.37	0.92	2.27	1.86
ATRIAS	50	2	0.42	60	3.90%	11	0.11	2	0.22
Delta Hopper	1	1	0.2	2	?	38	0.35	?	?
GOAT	1	1	0.26	2.5	25.20%	48	0.82	2.3	1.88
HRP3La-JSK	?	2	0.3	54	?	?	0.27	?	?
Minitaur	1	4	0.2	5	?	40	0.48	2.33	1.12
MIT Cheetah	5.8	4	0.275	33	10.30%	24	0.5	2.36	1.18
StarlETH	100	4	0.2	23	?	16	0.32	2.23	0.71
XRL	23	6	0.2	8	?	11	0.425	2.10	0.89
Salto	25	1	0.015	0.25	?	?	1.008	1.74	1.75

Table 8.2: Table comparing the different metrics of the robot developed in this work compared to other existing robots. The performance metrics for *BalekaBi* indicates the platform leaping with one leg and the other retracted while *BalekiBi2* metric are with both legs actuated. The body mass for the platforms in this work include the mass of the 1.42Kg boom. The leg mass (946g) percentage is a percentage of the robot body mass excluding the boom mass. Note that the max jump height is not an average but the highest ever reached by the robot [1][19].[36].

purely a result of the leg impedance and motor inertia opposing the back drive-ability. This was due to when the foot impacts the ground, it was stationary relative to the body and the motors inertia had to be overcome. If the motors and legs were massless, the legs would be back-driven without any resistance and no impact spike would be seen. The sudden spike was also contributed to the play in the legged system.

8.6 Robustness Investigation

Following significant testing, the robot was disassembled to investigate the robustness of parts. It was seen before disassembly that the play in the leg linkage had drastically increased from when the platform was first assembled. This is visualised in Figure 8.13 by the area in which the foot can move while the motors are held in place.

Upon opening up the transmission, it was noticed that the male gearbox interface shaft had very slightly yielded around the key slot. In addition, the key itself had started to yield. This can be seen in Figure 8.14 and 8.15. This increase in play was detrimental for any legged control system. Without encoders on the output, there was a large unknown region in which the end

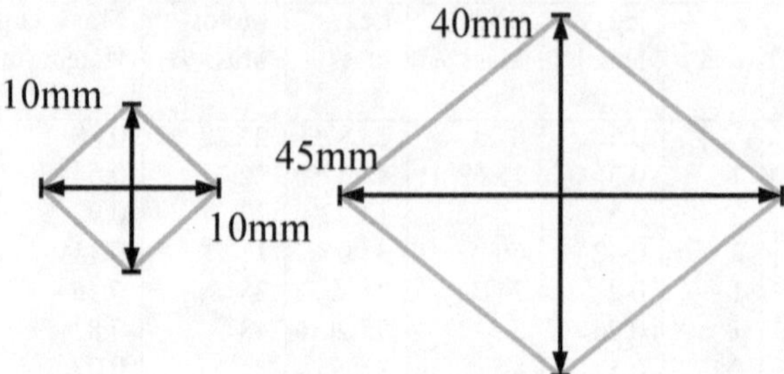

Figure 8.13: The left shape is made by the play in the end affector before testing takes place. The right shape is after all the experiments above have been conducted. The area of play increases by a factor of 18.

affecter could be.

Unfortunately, this plastic deformation could not be avoided. The company that supplied the gearboxes (Matex) were already specified at the start of the project, as mentioned in Section 1.4. During the design phase, it was decided to buy a drive shaft specifically made by the same company to interface with the gearbox that was needed. There was only a single shaft available to purchase and the author assumed that the shaft was designed adequately to match the planetary gear's torque rating. Furthermore, hand calculations and an FEA were performed to ensure the strength of the part (determined to have a FoS of 1.35) given the forces generated by the trajectory optimisation in Chapter 4.

However, as seen from the experiments, the shaft still plastically deformed around the key. The author suspects that due to the high reflected inertia of the motor the shaft experienced the majority of the GRF impact as a static load. Furthermore, the hand calculations used a fit factor that, if changed slightly, could decreased the FoS. The author select the factor according to detailed design tolerances, however, inaccuracies may have arisen from the workshop. In addition, the gearbox manufacturer did not offer any alternative shafts and it was suspected that the shaft would have the same rating as the supplied gearbox. Suggestions are provided in Chapter 9 for future designs of the platform that concern having a much higher FoS.

It was hoped that the robot would be able to make full use of the Raibert controller designed in Chapter 6. However, through several attempts, the biped was unable to remain in a stable limit cycle and tended to fall over. This was a direct result of the huge play that had occurred in the linkage mechanism. No other parts throughout the mechanical design had any visible damage or plastic deformation.

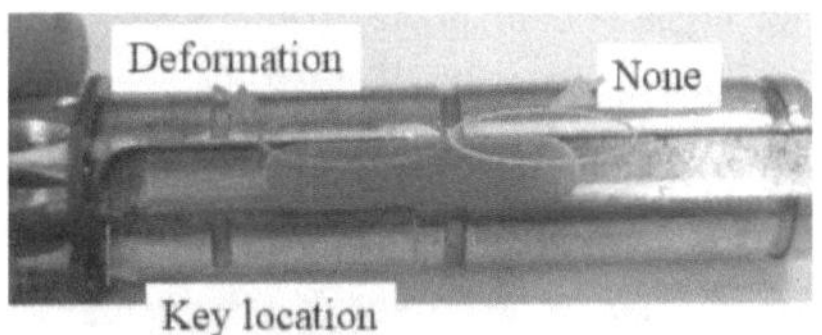

Figure 8.14: The deformation of the shaft around the key after numerous experiments

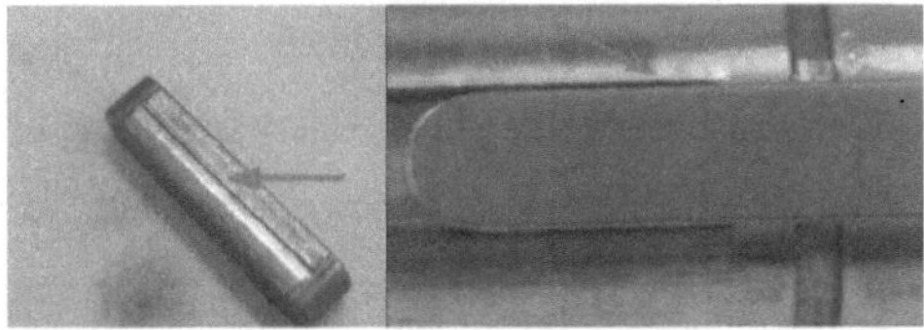

Figure 8.15: The deformation of the shaft key and the shaft.

8.7　Summary of Results

The purpose of this project was to develop a highly agile platform that would be capable of rapid acceleration manoeuvres. During initial testing on the prismatic support, it was identified that the U13 motors, even though have a greater torque constant than the U12 motors, were less adequate for this platform, suffering from severe cogging and a lower torque density. Thus, even though the U13 motors had been purchased, they were disregarded for the U12 motors. Furthermore, on the same rig, the original rubber damping foot was found to be too soft and compliant. This caused the robot to have excessive out of plane oscillations and the foot to bounce of the ground several times before settling. By replacing this foot with a steel cylinder and 3mm piece of rubber, the settling time was reduced by 50% and the foot only bounced off the ground once.

The force sensing capabilities of the platform were good with an average proprioceptive sensitivity error of only 16%. This was on par with other platforms such the GOAT leg and MIT cheetah having an accuracy of 19% and 5% respectively [19][27].

The vertical agility of Baleka leaping off one leg was found to be 1.33 m/s which was greater than that of a human's at 0.89 m/s. The biped leaping with both legs was also seen to outperform all other robots with a VA of 1.86 m/s, with the exception of the GOAT leg with a VA of 1.88

m/s, noting that it does not have to support the weight of a boom like Baleka does. Furthermore, Baleka was able to jump the highest out of all the existing platforms. Leading in vertical agility and hopping height indicates that this robot is the most agile biped making it suitable to perform rapid acceleration manoeuvres.

Lastly, the robustness was investigated and found that the main transmission shaft plastically deformed around the key. This was highly undesirable and introduced play into the leg linkage. Unfortunately, the platform failed to be robust enough against the high impacts of legged locomotion and this shaft needs to be redesigned. This was a result of poor shaft design by the manufacturing company for the gearbox not allowing the author to improve the FoS during the design phase. All other aspects of the mechanical design seemed unchanged after the experiments.

Chapter 9

Conclusion

The aim of this project was to build and test a bipedal platform, *Baleka*, that is suitable for rapid acceleration manoeuvres. This required a detailed design process as well as all the hardware and software systems to be developed for the robot's performance to be verified. To judge how the final robot performed, it is compared against other robots as well as by the requirements written up in Section 3.1.

Initially, the linkage topology and actuator scheme is investigated and selected to promote high powered, high speed motion. As seen from Chapter 4, a novel trajectory optimisation problem was then used to assist in the sizing of the bipedal platform. The optimisation problem was constrained by the selected linkage mechanism and pre-purchased motors. The solutions generated the optimal rapid acceleration motions, given these constraints. Furthermore, it assisted in identifying the most suitable leg length and gear ratio to use on the physical platform to have the fastest sprint. This was published and presented at the International Conference on Robotics and Automation (ICRA) 2018 [17]. This process resulted in the detailed mechanical design and assembly of the bipedal robot.

The mechanical design promoted a lightweight leg that weighed roughly 13% of the body weight, only slightly heavier than those seen in other platforms such as the MIT Cheetah. The leg design was modular so each leg could be repositioned and numerous bolt holes were added to allow other elements to be added such as an active tail if needed in the future. However, the option for adding in parallel or series elastic actuators if needed was avoided and should possibly be added in future iterations. The mechanical design (materials, machining, labour and material post processing) was kept as cheap as possible costing roughly R22 000 (US$1600).

Other systems that were set up within this project were a motion capture system for determining the robot's inertial frame position and velocities, the Simulink Real-time operating system, the EtherCAT communication protocol, other sensor integrations, a Raibert controller for the bipedal platform (simulated in a physics engine) and a boom to limit the robot's motion to the sagittal plane. This provided the supporting systems for the robot and any future iterations.

As seen from the results and discussions in Chapter 8, several key requirements were met. Baleka was shown to be the most agile robot with the highest vertical agility of 1.86 m/s and could also jump to a maximum height of 0.92 m when using both legs simultaneously. This level of agility is also greater than that of any human and existing robots. Thus, Baleka is considered suitable for rapid acceleration manoeuvres.

The need for a robust platform was not entirely met. The shafts provided by a supplying company did not design them with an adequate FoS for the interfacing planetary gears. The result was slight plastic deformation around the key slot, increasing the play of the leg linkage. This is not suitable for accurate foot positioning and was seen to be detrimental for the position control of the Raibert Controller. Improvements need to be made in the second iteration to overcome this shortfall.

To this end, the above shows that the platform design is agile and suitable for rapid acceleration manoeuvres. Furthermore the supporting systems are set up and ready for further experimentations and future robots.

www.ingramcontent.com/pod-product-compliance
Lightning Source LLC
LaVergne TN
LVHW091550170726
843492LV00007B/2128